Dale E Maddy

MILLION DOLLAR EXPERIENCE

April 8, 1966 – May 16, 1972

A Sailor's Viewpoint

D. E. Maddy

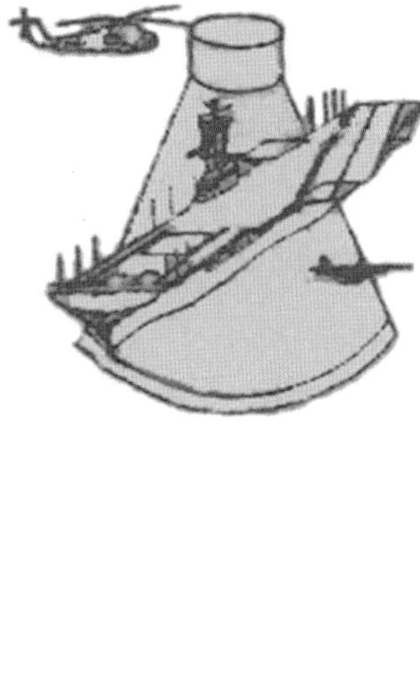

First Edition

DEM Graphics
Lodi, Wisconsin

ISBN: 978-0-615-97979-3

www.milliondollarexperience.net

Cover design by DEM Graphics

Cover photo by Roy Moruss

“If you've been in the military, reading MILLION DOLLAR EXPERIENCE will bring back wonderful memories (as it did for me). But if you haven't, you'll get an up-close look at the life of a US Navy Sailor. In either case, you'll be intrigued and inspired at the successful life of Dale Maddy and the down-to-earth, can-do attitude he honed while he was a "flat-top" sailor aboard the USS Kearsarge.”

Capt. Charlie Plumb
Carrier Pilot and POW '67-'73

“Dale, this is a very interesting book.”

Petty Officer, Leonard Joslin
USS Quincy survivor, WWII

“Holy cow! How great is this? To think a bunch of kids barely out of our gym shorts traveled that distance is amazing. If you think about it, we left the security of our family homes, found our way to boot camp, then to a ship, and then to foreign countries where we couldn't even speak the language. We all hung out together and took care of each other. I don't believe there is anywhere else that this goes on except in the military. You have a wealth of memories here that means a lot to the people that went through it. Can't tell you how much I appreciate you taking the time to put this all together. I know this must be a tremendous amount of work. I hope you enjoyed doing it because I sure enjoy looking at all of this.”

Seaman, Jack White
USS Kearsarge

This book is dedicated to…

All the men and women who
Served in the United States military

Especially to those who gave
The ultimate sacrifice, their Life

In memory of

1st LT Larry R. Maddy US Army
LTJG Ernest J. Krajsa USNR
LTJG Richard Sullivan USNR
AX1 Thomas Goen USN
AN Robert Yoder USN
SFM3 Harry Hope USN
AK1 Kenneth Edwards USN
SA Kenneth Barslow USN
FN Rodney Shipman USN
The 77 men of the Frank E. Evans who died at sea
CDR Bruce Simonds USN
CDR Denny Phillips USN
LTJG Donald Hugo Hagge USN

The Maddy Brothers

SGT. Wayne G. Maddy, US Air Force – Argentina/Panama
ETN3 Dale E. Maddy, US Navy – USS Kearsarge Vietnam
1st LT. Larry R. Maddy, US Army – KIA in Vietnam

SPC4 John M. Maddy, US Army – Ethiopia
CPL. Paul G. Maddy, US Marine Corp. – Okinawa
Joel Maddy – too young to serve

Special Thanks

Special thanks to all of those who helped me with this book. To those who proofread my manuscript; Barbara Carper, Marvel Maddy, Rob Bryant, and Alicia Raisbeck. Especially to Sandy Maddy who makes me look like I am a better writer than I really am. Thanks to Joel Maddy for writing and marketing ideas. Thanks to the few that encouraged me to continue when I had doubts. Thanks to those who helped clarify story details, Sonny Richardson, Barry Rittle, Ron Brown, John Hand, Steve Spratt, Lynn VanAcker, Jack White, Ron Smith, and Leonard Joslin. Thanks to my daughter Tina for her help with photographs and cover design.

Preface

We go through life having many experiences just from everyday living. From those experiences, we have many stories to tell. Some of those stories are interesting and some maybe not so much. Some are humorous, some are sad, and some are hard to believe. When we die, those stories die with us – that is unless we have written them down. Then they can live on for hundreds of years after we have long been forgotten. It is for this reason I have written Million Dollar Experience. If this book is enjoyed by just one person, the time spent writing it will be worth it. These are my experiences and the people with whom I shared them. Those experiences are my life. They influenced who I have become and how I see the world. Sometimes when younger people know what has happened before them, they can make better decisions about their own lives and make even better things happen. This book is the result of organizing my thoughts and memories, from my Navy days and things related to it; and from both before and after those days. After 47 years, I still remember all these stories.

There were about 4000 men on the USS Kearsarge and I am sure each one has his own story to tell. If two men witness the same incident, there are two stories that may not match. Each story is relevant to his station or to what he was told or learned from someone else, or experienced firsthand.

The stories are from my perspective and not necessarily the official record of the United States Navy. Some of the names have been changed to protect the innocent or the guilty.

Every time in my life when I would start a business, people would say, "Are you kidding, you'll never make it." Usually the people who are closest to you say that. Well the same thing happened when I said I was going to write this book. They said, "Who cares about you and your life; you are not famous." The answer to that is you do because you are reading it right now.

When I was in sales, we used to have a saying that if you were wearing a tie and more than fifty miles from home, you were an expert. It is like the ancient Chinese secret ingredient for losing weight or for staying young. People think that because it is from a faraway place and is kept secretive, it has mystical power. That is why in the 1890's, the traveling medicine man got away with selling his snake oil. People think you cannot write a book unless you live in New York, L.A., or some other faraway place.

Well, unlike Nancy Pelosi's Health Care bill, you will have to read this to find out what is in it. Then you decide if you enjoyed it.

Table Of Contents

Chapter 1 - Childhood Days

War and Dying

I was born in a farmhouse in Belleville, Ohio on a stormy night in May 1947. World War II was over and the Korean War had not started. It was a time of peace in our nation.

Dale, Larry, Wayne - 1952

Before I reached the age of one, my family moved to Granville, Illinois. During my pre-school days, we lived in a small two-story house across the street from the grade school. One mid-afternoon in 1952 when I was five years old, I was sitting on the couch with my mother. She was taking a break from her daily chores. I was not in school yet, but my two older brothers and two older sisters were. My younger sister and two younger brothers were taking a nap. It was just the two of us. As she sat there, she started thumbing through a copy of *Life* magazine. I sat closer so I could see the pictures of soldiers in the Korean War. At the age of five, I asked many questions. There was a picture of the explosion of a hydrogen bomb (the United States had just exploded its first hydrogen bomb in November). She said this was a hydrogen bomb and not an atomic bomb. I really did not know the difference and my mom probably did not either. I wanted to know what it was. She tried to explain how bad the atomic bomb was, and that it should never be used again. She said it made America strong and helped win the war against Japan. I wanted to know what war was. She said war was bad and that she hoped her boys would never have to go to war. She said that WWI was the war to end all wars, or so said President Woodrow Wilson. Now here we were, still at war. She said that many people were killed and that was bad. At this point in my life, I did not know what it meant to be killed or to die. My mom said, "It's like going to sleep and never waking up again." I could not relate to this because I did not know anyone who had died. I did not understand most of it, but I also never forgot what she said. I felt that if WWI was the war to end all wars, that at least I was safe because there never would be another war. This conversation was a very memorable one for me.

I did not know anything about war at this point in my life. I had a neighbor, Wayne Sandburg, who was about my age, who seemed to know quite a lot about it. The Sandburg family had a TV, but the Maddys did not, and I had not been exposed to all those things. Wayne had been exposed to all this stuff. He had the full chest of toys, the little green toy soldiers, like the ones in Toy Story, cap guns, an air gun that shot ping pong balls, and toy airplanes, cars, trucks, etc. At the age of six, Wayne and I were playing in his sand box. It was the largest sand box I had ever seen. It was two feet deep and eight feet by eight feet square. Wayne picked up a toy airplane and started saying, "Bombs over Tokyo." I had never heard that saying before and did not know where Tokyo was or why it was the town to bomb, but it sounded good. I thought the name Tokyo had a nice sound to it. There must be a good reason for Tokyo to be the chosen town to bomb. I did not know what it was, but I was willing to go along with it. In the many times I spent with Wayne, we played cowboys and Indians or cops and robbers, or US Army and Japs. I did not know what japs were either. I soon learned the good guys versus bad guys concept. Wayne always wanted to be the good guy and I had to be the bad guy. They were his cap guns, so I either went along, or went home.

During the time between 1954 and 1957, I rarely went to a movie theater, but when I did, it was only to see Roy Rogers, Abbott and Costello, or stories from the Bible, such as "Samson" or "The Robe." It was in the Bible story movies that I realized what death truly was.

A couple of years later, my Great Grandmother Throckmorten passed away, and it was just as my mom had described, she went to sleep, never to wake up again. Grandma used to make me and my siblings kneel at the bedside and say our prayers before going to bed. I never forgot that prayer...

> *"Now I lay me down to sleep, I pray to Lord my soul to keep, if I should die before I wake I pray to Lord my soul to take."*

Living on the Farm

In 1954, we moved to a farmhouse a mile east of Granville, Illinois. We played in the grain bin, built tunnels in the haymow, and set fire to the pasture. Yeah, we pretended we were cowboys burning the Indians off the land by setting it on fire. Do movies influence kids? Yes!

On the farm, at eight years old, I scooped manure, worked a garden, and pulled weeds in the fields. I cut thistles that were taller than I was for a penny per thistle. At the age of nine, I was mowing lawns to earn money at 50 cents per lawn.

About two hundred yards south of the house, in the middle of the pasture, was a pond. It was a source of drinking water for the cattle. The pond, about 60 feet in diameter, was manmade. One side of the pond had a sandy beach area and was about two feet deep. The other side was about ten feet deep with a muddy bottom. In the winter; the pond would freeze and we would ice skate on it. In the summer we would play on the sandy beach.

My brothers Wayne and Larry built a raft from old fence boards. We put it in the pond and pushed it around with a pole like Huckleberry Finn. One summer afternoon the Tentacle brothers, Tony and Billy, had come from town to play with us. Billy, the older of the two brothers, took the raft out for a ride. When he pushed the pole down on the muddy side of the pond, it stuck in the mud. Instead of letting go of the pole, he held on to it and the raft floated away. The pole started to lean and a second later this kid was in the pond, on the deep side. His younger brother started yelling, "He can't swim." My brother Wayne swam to his rescue. When Wayne got near him, he pushed Wayne under to get himself up. Wayne managed to get him close enough to shore that we could pull both of them ashore with a long pole. At the ripe old age of ten, Wayne had already saved someone's life.

It was in this pond that I taught myself how to swim. Swimming just came naturally. My brothers could do it, and I just mimicked what I watched them do.

Small Town America

By the time I was ten years old, my dad made a career change and moved the family to Sheffield, Illinois. Sheffield was a small, quiet town of about 1000 people. It was a great place for a young boy to grow up. No one locked his doors at night. The high school had only about 250 students from the freshmen through the senior class. Everybody knew everyone else. I could carry my rifle down Main Street and nobody gave that a second thought. Kids did not kill kids. You could take a pocketknife to school and no one cared. You could talk about guns and no one cared about that either. It was more horrible to have a copy of *Playboy* magazine in your locker than to talk about guns. The boys all joined the Boy Scouts and the girls joined Girl Scouts. Almost everyone went to a church of his or her choice. Everyone was Caucasian of European decent except for one Mexican family. The Rameras family had a boy, Daniel, who was one year older than I was. His brother, Paul, was a year younger than I was. They were both good kids. I became friends with them. Some of the older folks could not speak English, but no one cared about that either.

We recited the Pledge of Allegiance every morning in school…

> *"I pledge allegiance to the Flag of the United States of America, and to the Republic for which it stands, one Nation under God, indivisible, with liberty and justice for all."*[1]

We recited the Lord's Prayer in church every Sunday…

> *"Our Father, which art in heaven, Hallowed be thy Name. Thy Kingdom come, Thy will be done in earth, as it is in heaven. Give us this day our daily b. And forgive us our trespasses, as we forgive them that trespass against us. And lead us not into temptation, but deliver us from evil. For thine is the kingdom, the power, and the glory, Forever and ever. Amen."* [2]

Everyone sang the first verse of the National Anthem at sporting events…

> *Oh, say, can you see, by the dawn's early light, what so proudly we hailed at the twilight's last gleaming?*
> *Whose broad stripes and bright stars, thro' the perilous fight, o'er the ramparts we watched, were so gallantly streaming?*
> *And the rockets' red glare, the bombs bursting in air,*
> *Gave proof thro' the night that our flag was still there.*
> *O say, does that Star - Spangled Banner yet wave o'er the land of the free and the home of the brave?*

In our school, we were taught American History and the United States Constitution. The kids respected their parents and said, "Please" and "Thank You." We called older men "Sir" and older women "Ma'am." The kids did not smoke dope or take illegal drugs. However, some did smoke cigarettes and occasionally drink alcohol if they could get it. I never participated in either of those things. The worst mischief the boys did was throw garbage in the basketball coach's front yard late at night. On the Fourth of July, there were always big celebrations and many fireworks. People were friendly and welcomed outsiders with open arms. In the summer, we played Kick-the-Can and Hide-and-Seek and caught fireflies. We watched as Sputnik flew over our house. We would lie on our backs in the front yard and look at the stars, spotting the Big and Little Dipper. I have many great memories of those days. That is the Great America that I remember and the one I would fight to defend. We are as American as it gets.

BB Gun Incident

In 1960, I had purchased a BB gun and had become a very good shot. I could shoot a pigeon in the head from about 50 feet away. We sold them for ten cents each to Mrs. De Marley, a Belgium woman who ate them.

One day my older brother Larry took my gun away and kept it. Larry was that way. He would sometimes antagonize just for the sake of fighting. I wrestled with him to get it back. He then started to come at me vigorously. I cocked the gun and told him if he took one more step I would shoot him in the eye. He did not believe me and started to lunge at me. From about 20 feet, I shot him and hit him on the right cheek, about two inches below his eye. That was a warning shot, the next one would be in his eye. I did not aim at his eye or I would have hit it and then I would have been in big trouble. It broke the skin and started to bleed. This really pissed him off and he grabbed my gun and broke it in half over his knee. Then he threatened to kill me and came at me swinging the broken gun. Larry was an inch taller than me and about 20 pounds heavier. Overall, he was bigger and stronger. I needed an equalizer. I saw a hammer lying on the porch, so I picked it up and stood my ground, swinging the hammer with all my might. He took my gun, ran into the house, and locked the door. I told him to unlock the door or I would bust it with the hammer. He did not take me seriously, so I slammed into the door with the hammer and put a big hole in it. Larry then unlocked the door and walked away, leaving my broken gun behind. It was all over.

Now, I was in big trouble for breaking the door. My dad was not very pleased with me and reminded me of that broken door many times after that. Somehow, Larry escaped any of the blame even though he started the whole thing. Nothing was ever said to him for breaking my gun. This incident is not something I am proud of, but it did happen, and it is relevant to another story in this book.

Losing a Friend

In the spring of 1962, I had a schoolmate named Bob Martin. Bob and I had some fun times together at school functions. There were times that Bob was a little mischievous. One night while at the high school for an event, Bob, Cary Standard, and I were messing around in the school maintenance shop when Bob decided he wanted to steal the anvil. An anvil weighs about 130 pounds and is hard to carry. The anvil weighed more than anyone of us did.

I said, “Why do you want that?”

Bob said, "Because it’s here and I want it.”

I said, “I will have nothing to do with stealing.”

He told me to watch the hall to make sure no one was around. While standing by the door, I could hear Bob and Cary grunting and talking. After about ten minutes, he called me over and they had only moved it about six feet.

Bob said, “It’s too heavy, I guess that was a bad idea.”

Therefore, we left the anvil behind and exited the school. I always wondered what the school maintenance man thought when he found the anvil had been moved six feet.

In 1963, at 16 years old, Bob Martin died from injuries received in a car – train crash. It was a sad day for everyone in the community and in the high school. As bad as his death was, it brought home the reality of death and that it can happen to any of us, even at a young age. This experience taught my friends and I that life is precious and we should not take stupid risks or we could be next.

Being Responsible

At the ages of 14 and 15, I worked for the Market Basket Grocery store on weekends. The store was owned by Earl Radcliff, an old man who went by the name of Fuzzy. I do not know why everyone called him Fuzzy, but so did I. Fuzzy was an alcoholic, and I remember him having a hangover a great deal of the time.

One Sunday afternoon I was working the store. I was there by myself. Larry and about ten of his friends came into the store and started stealing candy. Their pockets were bulging. They were circling me like vultures and I was not sure what to do. There were so many of them I could not watch them all. Fuzzy was in his second floor apartment with a hangover, so I knew I could not count on him. There was a bell switch by the front door that I turned on. It sounded like a burglar alarm. When they heard it ringing, they scattered like a bunch of scared rats, right out the front door. I never called the police because these were some of my friends, too.

Fuzzy owned a 1954 Chevy station wagon that he used for home deliveries of groceries and taking boxes to the city dump. Because he was often drunk, he rarely drove his car. He lived in the store and the bar was only a couple of doors up the street. Even though I was only 15 and did not have a driver's license, he would let me drive his car to make those home deliveries around town and trips to the dump. I never went anywhere else.

In the summer of 1962, my parents were driving to Chariton, Iowa to visit my grandparents for the weekend. Larry and I stayed home because Larry had newspapers to deliver, and I worked at the Market Basket. On that weekend, Larry convinced me to ask Fuzzy if we could borrow his car so he could make his newspaper delivery on Sunday morning. Fuzzy would only let us have the car if Larry were the driver because he was licensed and I was not.

That night Larry decided he was going to use this car to go see his girlfriend about ten miles away. Actually, that was his intention from the beginning, he just did not tell me. She lived on a farm and he could not find it. He was unfamiliar with the roads. He came into a T intersection while driving too fast and rolled that car into a ditch. The crash broke the windows and smashed the top. A tow truck pulled us out of the ditch, but no police were ever called. So Larry did not get a ticket. The car was still drivable, so Larry took it back to Fuzzy and walked away.

Fuzzy told me I would have to work for no pay until the damages on his car were paid. I worked for a year at 50 cents per hour to pay off the damage that Larry had created. I paid about $600 in total. When I asked Larry to help pay, he refused. I thought, someday he is going to pay me for his share of that damage. Just for the record, Fuzzy never did repair the car.

Draft Registration

In 1965 when I was a senior in high school, the word among my friends was that when you turn eighteen, you must register with the Selective Service (draft board). In May 1965, just four days after my eighteenth birthday, I drove to the Selective Service Office in Princeton, Illinois. It was located in the post office building, and while there, I signed my name to serve the United States government. I had my focus on my upcoming electronics school in Rock Island, Illinois, not on war. I signed up, not because I wanted to serve, but because it was the thing to do. I did it mostly because it was against the law not to. I wanted nothing to do with the military and war. When I was a teenager, I knew people who had gotten into trouble with the law, and they were sent to the Army rather than going to jail. That surely did not send a very good message to people like me. I interpreted that to mean that being in the Army was like being in jail; it was a punishment.

Two weeks later I enrolled in RETS Radio Electronics Television school in Rock Island, Illinois, to learn radio and television repair. I attended a four-hour class one night a week. While at RETS, I built an AM radio from scratch, learning one circuit at a time.

One week later, I graduated from Western High School, in Buda, Illinois. I always said I was the one who made the top half of the class possible by being in the bottom half. I was what you would call an average student. I received mostly C's and occasionally a B, with an A from time to time. My favorite subjects were math, history, speech, and electricity shop. My worst classes were English, biology, and science.

When Bob Martin died, Readus Quickle offered me the job that Bob had been doing. I accepted the offer and worked for Quickle Farm Implement Company during my junior and senior year. I continued to work there for a couple of months after graduation. During these years my dad told me always to do my job as well as I could, and to do more than is expected. That will always help you get ahead. He said, "Somebody will notice your good work, even if it's not the guy you are working for." The job was a handyman and a "goffer."

During the day, the radio would play in the shop and the Vietnam War was constantly in the news. At the time, President Lyndon Johnson was going to make an executive order that would exempt from the draft any man who was married before midnight August 26, 1965. The shop mechanic teased me about getting married to avoid the draft. I did not think getting married just to avoid the draft was a good idea. Besides that, I was not seriously dating anyone at the time.

Prior to 1965, I did not pay much attention to world events. It did not mean anything to me. I did not even know where Vietnam was or why it had anything to do with my life in Illinois. My feeling was that if people in other countries want to fight, go for it, but leave me out of it. On August 5, the Selective Service notified me my draft status was 1-A, which means available for service.

I continued to live at home with my parents while I was attending RETS school. I felt the goffer job I held at Quickle Implements was great as a part time job while in high school, but now I needed to make more than $1.25 per hour. In the middle of July, I quit working for Readus Quickle, and started working for Barcol Overdoors as a rail bander and increased my income to $2.50 per hour. On September 1, I accepted a job offer at John Deere Parts Depot in East Moline, Illinois as a parts picker filling orders for John Deere dealers in the Iowa and Illinois area. This worked out well for me because now I would already be in the Quad Cities on Tuesday nights for my RETS school. This new job paid me a whopping $3.25 per hour. This was a good hourly wage at that time in history. Co-workers were supporting families on that income.

John Deere Parts Depot

All was well for me; I was working a 40-hour week, and going to school at night and doing well in my studies. I had weekends to spend with my friends, was dating different girls, and saving some money too. I was anxious to get my school finished and get on with my life. I worked harder at getting good grades at RETS than I ever did while in high school. I bought all the components and built my own test equipment (volt/ohm meter).

Beating the Draft

On March 14, 1966, the Selective Service ordered me to report to Chicago for a physical exam. I boarded the Burlington Railroad's Denver Zephyr train in Princeton, Illinois, where I met two other men from Illinois who were going to the same place. Robert Courtney was from Canton and Craig Rigg was from Galesburg.

It was questionable as to whether I was qualified for military duty because I had above average blood pressure. The government doctors said it was just anxiety and passed me through anyway. Even though I had above average blood pressure five years earlier during a grade school sports exam, they still passed me. They just wanted bodies because Lyndon Johnson was escalating the war, and it required 35,000 new men every month in order to meet the quota demand of the generals.

It is hard to believe that in just twelve years I went from being a naive little boy playing in a sand box to being a man in a military uniform going off to war. It is no wonder that so many moms in America still think of their nineteen-year-old men as their little boy. I have heard many women say, "He is just a boy," when their son is 21 or 22 years old. By the time they get to the age which they can make their own intelligent decisions and have their own identity, our country sends them away. It just does not seem right.

While staying at the YMCA in Chicago, I met a young woman named Barbara Hoover who was staying at the same place. She was there on a school field trip, and I was there for my military physical. She was from Des Moines, Iowa. We talked briefly and exchanged names and addresses. I felt that I would probably never hear from her again.

My friends received a draft notice to the Army within 30 days after their physical exam. Knowing this, I was considering enlisting in the Navy to beat the draft. My RETS schooling did not disqualify me from being drafted, and I had already gone past the marriage deadline. I decided that if I had to serve in the military, it ought to, at least, be on my own terms. Since my brother Wayne was already in the Air force, I thought I would do something different. The Navy was advertising, "Join the Navy, and see the world," and that sounded like what I wanted to do. On April 1 the Navy recruiter, Chief Petty Officer Robert Bachman, sent me a letter announcing a new Cache program the Navy was doing. It allowed a person to sign up for the Navy Reserve now and go active 120 days later. I asked if the Navy would continue my education in electronics since I had already been doing that. Mr. Bachman said that if I passed the Navy test, I would be able to continue my electronic education. With that information, I quit attending RETS which was 65% completed at the time. I enlisted in the Navy seven days later. I asked the recruiter if he could get my draft status, and he said he could. A week later, Chief Petty Officer Bachman called me and told me my name was on the top of the draft list and they would remove it now that I had enlisted in the Navy. I would have been drafted into the Army within a week had I not enlisted.

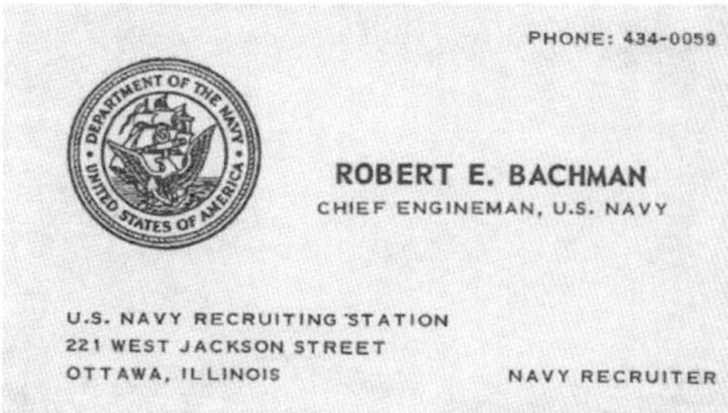

On May 8, 1966, the Selective Service required me to fill out an information sheet about my education and work history. It was just like a job application from any company. About a month later, I received a new draft card with a 1-D classification, which means I couldn't be drafted because I was already serving the government as a member of the Navy Reserve.

On August 8, 1966, John Deere granted me, in writing, a military leave of absence for the next four years. The company promised me my job back with equal or better pay when I returned. The next day I received a letter from the Navy Department telling me to report for active duty in three weeks.

During the next 21 days, I loaned my 33-RPM stereo record player to my sister Becky. I packed away my reel-to-reel tape recorder. I sold my Honda 250 motorcycle to my brother Larry. I sold my 1958 Studebaker Silver Hawk to a fellow employee at John Deere. The rest of my be-

longings (clothes, photographs, and high school memorabilia) were put in a footlocker trunk in the attic of my parents' home. At nineteen years old, that was everything I owned, not much.

Chapter 2 - Boot Camp Experience

Arriving at Camp Barry

The big day, August 30, 1966, Lynn VanAcker and I reported to Great Lakes Naval Center for boot camp training. Lynn was a long time high school friend of mine who lived across the street from me. We had enlisted together on the buddy plan under the Navy Cache program. The buddy plan, we were told, was a system in which you would be with a friend through your years in the Navy. We were told about the buddy plan by Chief Petty Officer Robert Bachman, the recruiter.

When we arrived at Great Lakes, we were fed supper, given bedding, and shown our barracks. The next morning we were awakened at 05:00. We went to breakfast by 06:00 and then on to start our two day "IN PROCESSING." After breakfast, we were divided into companies of about 80 men each. At this time, Lynn was sent to company 537 and I was sent to company 536. I spoke up and told them I was on the buddy plan and that Lynn and I were not to be separated. The First Class Petty Officer, W. S. Meiss, looked at me and said, "We don't have any such thing." This was the first lie I experienced from the Navy. I never saw Lynn again until we were home on leave. However, I did know a couple of people because Craig Rigg and Robert Courtney ended up in my company.

Taking an Oath

Next, we were taken into a large room to pledge an oath. The room was an official looking room with nice blue drapes in the front, wood paneling on the walls, and a Navy flag in one corner and an American flag in the other corner. An officer in his dress uniform administered the oath. It was all very official. The oath is much the same as that of the President of the United States, pledging to uphold the Constitution of the United Sates.

"I, Dale Maddy, do solemnly swear that I will support and defend the Constitution of the United States

against all enemies, foreign and domestic; that I will bear true faith and allegiance to the same; and that I will obey the orders of the President of the United States and the orders of the officers appointed over me, according to regulations and the Uniform Code of Military Justice. So help me God." [1]

There is that one part about obeying orders of the President with which I had a problem. Lyndon Johnson was President at the time. I did not agree with Johnson on his policies, therefore; I would not swear to honor him and his orders. The Navy did not care. By the mere fact that I was standing there with my right hand in the air it was affirmative. I did not even recite those words, and I was rammed through the system anyhow. I did not really want to be in the military, but on the other hand, I did not want to be sent home and be a protester or draft dodger. There is no honor in that.

In Processing

After swearing in, we were sent to the barber and all of our hair was shaved from our heads. Some of the men who looked like tough guys or a little freaky before did not look so much that way after. Now everyone was equal. The next stop was getting military issued clothes. The first piece of clothing we received was a bathing suit. We were told to strip naked, and after a naked body inspection, we put on our swimsuits. Our civilian clothes including watches, cameras, and money were then to be put into a box, addressed and sent home. I did not comply 100% to that because I kept my watch, my camera, and my wallet. Then everyone was given an empty sea bag (duffle bag) and our Navy clothing was issued, one piece at a time. First on the list were skivvies (boxer shorts), then tee shirts, then dungarees and blue Navy work shirts. First, I was issued my white uniforms and then blue uniforms. In addition, they gave us dress shoes, work shoes, hats, and winter clothes (the famous P-coat). The Navy thought of everything; we were given a personal kit that included a sewing kit, soap, toothpaste, toothbrush, etc. When the sea bag was full, everyone went into another room with rows of tables. Everyone was given a stencil with his name and serial number and instructed on how to stencil their name on every article of clothing.

A "Blue Jackets Manual" was included among the things issued to us. It was originally written in 1902 as a source of practical information about our Navy's ships and Navy living.

The next day everyone was taken into a room and instructed on brushing their teeth. *Yes, you heard it, brushing your teeth. Unbelievably, some people had never had any dental hygiene lessons in their family or in their upbringing.* Then everyone received a dental exam, and those who needed anything done were scheduled for dental work in the following days. Following the dental exams were those famous shots with an air pressure gun. We were vaccinated for everything under the sun and tested for viral infections

When IN PROCESSING was finished, we were moved to a more permanent barracks where we would live for the next nine weeks. There was some marching and carrying a rifle around, but the Navy training is different from the Army. Marching, shooting, and physical drills are not the focus of the Navy training. During the next few weeks, I was taught how to live in tight quarters on board a ship and how to care for personal hygiene. We were also taught how to swim, how to fight fires on a ship, and about ship's ordinance. We experienced tear gas, radio communications, and basic ship life. We learned what to do if captured by the enemy, and how to obey orders.

We were told that we could design a company flag with anything we wanted. Some of the guys thought it would be great idea to have Playboy send a flag. They sent Playboy a letter with their request, but Hugh Heffner turned them down. I am not sure why he would make a decision not to support our troops, but he did. Those same men purchased many of his magazines over the next few years. Everyone gave up on the idea. We did not have a special company flag.

Navy Basics

Some of the Navy basics are about a ship. First, what is the difference between a ship and a boat? They both float. Some have said it is a boat if it is less than a 150 feet in length and ship if it is more than that. The Navy says a boat can be loaded onto a ship. Submarines are called boats even if they do not fit either of those descriptions. We were taught ship terms, such as, the floor is the deck, the wall is a bulkhead, and the ceiling is the overhead. A hallway is a passageway, a stairway is a ladder, a window is a port, and a door is a hatch. A chimney is a stack. The front of the ship is the stem and the rear is the stern. That is where the term "stem to stern" comes from. When standing on the deck looking to the stem, the right side is the starboard side and the left side is the port side. The kitchen is the galley and the bathroom is the head. A sailor does not get out of bed and go to work; he hits the deck and turns to. He does not mop the floor; he swabs the deck. A sailor gets his sea legs from walking on a deck that is constantly rocking and rolling. It did not take long to learn the terms because that is the way everyone talks and I picked it up right away.[2]

Ships Around the World

The U.S. First Fleet is made up of ships sailing from, or with homeports off, the West coast to Hawaii in the Pacific Ocean. The U.S. Seventh Fleet is ships sailing from Hawaii to the Far East. The U.S. Second Fleet is ships sailing from, or with homeports off, the East coast in the Atlantic Ocean. The U.S. Sixth Fleet is ships sailing in European waters and the Mediterranean. The U.S. Mid-East Fleet is ships sailing east of South Africa in and around the Indian Ocean. The Navy operates in all oceans of the world. When a show of force is needed to bring peace, the U.S. Navy can be anywhere in the world in a very short time. Ships are already in many locations on a daily basis. Ships stop in ports around the world, but there is never a time when every ship is in a port. Our submarines can be anywhere, and no one knows it. They can pop up anytime, anywhere.

Swimming

During the swimming lessons, sailors were taught how to survive in water with oil fires on the surface. When a ship explodes, there is bound to be lots of fuel on the surface of the water, and, most likely, it will be on fire. If you are swimming near it or in the middle of it, all you need to do is swish your arms from side to side, and it will make the oil flow away from you and the fire will follow it. Knowing how to do this could save your life.

We were also taught how to use the Dixie Cup hat or our dungarees as a floating device. Taking our Dixie Cup hat and turning it into an upside down bowl and keeping it wet, it will hold air. By keeping it out in front of you while on your stomach, you can float with it indefinitely. If you tie the end of each of your pant legs into a knot and keep them wet, they will hold air also, and you can lie between the legs with them in the air. If the air starts to seep out just take a deep breath, go under the water, and blow air back in the leg. The pant must be kept wet at all times to hold air.

We were taught how to jump into the water from a ship. The ship you are stationed on will determine the height from which you would have to jump. If the ship is sinking or if it is burning, you may have to abandon ship and get in the water. Always jump feet first, never dive. There may be debris in the water and diving could crush your skull or break your neck if you hit something. If you have an inflatable life vest on, do not inflate until you are in the water. When jumping, cross your arms and hold your nose so water does not go up your nose. You may think you do not need to do this but you do not want water with oil and gas going up your nose. Remember, the water is salt water and you do not drink it.

I thought the drill instructor was going to drown the black men. The pool was an Olympic size and the drill was to swim the length twice and then tread water in the deep end for ten minutes before getting out of the pool. Those black men kept going under and swallowing water. It looked as if they were on the edge of drowning just as the instructor plucked them from the water. There were only five black men in my company of 80, and they were the only ones who could not swim. I thought it was strange that they could not swim. I felt sorry for them and the abuse they were experiencing. I was told later that because of segregation they were never allowed in public swimming pools. I thought this was strange, too, because I taught myself to swim in the pond on the farm in Granville, Illinois. Maybe those black city kids from the south just did not have a pond or a river in which they could swim. If you could not swim, you were sent to a special swimming class. The rule in the Navy: if you could not swim, you were sent home. No one wanted to be sent home as a loser. One of those men was sent home because he just could not grasp swimming. That may sound like a good thing for someone who wants out of the Navy, but he was probably drafted into the Army as soon as he arrived home.

Fire Fighting

Firefighting school was an interesting and a somewhat fun class. The men were taught how to fight oil and gas fires and how to maneuver through the ships' corridors in the dark. We were taught how to go up and down different levels while in black smoke. Put your hand near a bulkhead, and if it is hot, you have a fire on the other side. Do not go into that compartment. Sometimes buttoning down the hatch and starving the fire from oxygen will extinguish a fire. Fighting fires is a scenario that has been real for many a sailor over the years of Navy history. It is necessary that every sailor learn this information. It is conceivable that within weeks after leaving boot camp, a sailor could find himself in this predicament.

Nothing Lasts Forever

When we were marching and carrying our rifles, we were asked to hold them straight out in front of ourselves. Out of curiosity, I timed it to see how long we would have to do this exercise *(Remember I kept my watch)*. To some of the recruits it must have seemed like an eternity because they were starting to whine about their shoulders hurting. The truth was: what seemed like ten minutes was only about 60 seconds. What I learned from that was, "Nothing in life lasts forever and we should try to endure the pains for it won't be forever. Also, things of joy don't last forever and we should savor them when possible."

K P Duty

By the sixth week of training, everyone was sent to "service week." During service week, everyone is assigned a job serving the rest of the recruits. My job was in the galley as a breakfast cook. I was to cook scrambled eggs every morning. On the previous day, cases of eggs were brought into the kitchen and we were to break the shells and fill two tubs with the eggs. There must have been sixty gallons of eggs in those tubs. When we were finished, the tub was covered and wheeled into a large refrigerated room. The next morning when I was cooking the eggs, people were complaining that shells were in the eggs. When those eggs were being broken, some of the sailors were showing off by breaking four eggs at a time and some shells had dropped into the tub. The chief dumped both vats down the drain. What a waste! I tried to convince him I could find and remove the shells to save those eggs, but he said, "No." There were no eggs for breakfast that morning.

The next day we were taken off the egg detail and put on an oven cleaning detail. The oven was eight feet wide and eight feet tall. We were to get inside and scrub it down with a wire brush. This took us several days to accomplish. The officer said we had done such a fine job, he was going to put a letter of commendation in our service record, but he never did.

The Gas Mask

The day everyone was learning about the gas mask, we were taken into a building and given a mask and instructions on how to use it. I had received a mask that had not been thoroughly cleaned from the previous class, and there was gas residue inside the mask. When the trainer instructed everyone to put on the mask, I was in trouble immediately. I was not allowed to take the mask off my face. Besides, the minute everyone had a mask on, he ignited the tear gas canister. I could not win on this one as I had gas in my mask and gas outside the mask. The final move of the training was taking the mask off just before leaving the building. I was gagging so badly with the mask on that I thought I would be better off with the mask off, so I ripped it off my head. It was horrendous; the gas went into my eyes, up my nose, and down my throat. It was all I could do to get out of that building. I could barely breathe or see where I was going. I coughed and my nose started running endlessly, and my eyes burned and were red and tearing. All I could think about was getting as far away from the building as possible. Years later when I heard war protestors talking about being gassed, I knew what it was like.

Navy Nonsense

One of the things that really irritated me about military nonsense was the rules for using the bathroom facilities. There were about ten toilets, 20 sinks, and ten urinals for about 80 men. That should have been an adequate number, but some higher-ranking person decided that the men should use only two sinks, two toilets, and two urinals. The men were told that if we did not use the rest of them, they would always look clean and when inspection day came, it would be easier to pass inspection. It was true, eighteen of the sinks, eight of the toilets and eight of the urinals were in brand new condition.

One day when the company returned from a day of marching, everyone ran to the head to pee. There was a line of 80 men waiting to go. One of the men had to go bad enough that he passed everyone and used one of the forbidden toilets. He was put on report and his punishment was to clean the head with a toothbrush for the next week. This is why I disliked the military. It was the most ridiculous idea I had ever seen in my life. I was told years later that someone had written to their congressional representative about this and they stopped doing it.

A Thief

When we washed our clothes, we hung them out to dry on a clothesline behind the barracks. In November, the temperature at Great Lakes Naval Base was below 30 degrees. When we went to retrieve our clothes, they were frozen solid. When one of my friends and I went out to retrieve our sheets from the line, they were missing. (*When we arrived at boot camp, everyone was issued two sheets, one pillowcase, and one blanket. We were told that is what we must return when we leave or we would be billed for it, and it would be taken out of our pay.*) For some reason, someone saw fit to steal my sheet, so the only recourse I had was to steal someone else's. After it was dark, my friend and I went to the company next-door's clothesline and picked off a couple of sheets. Just as we turned around someone yelled from the third floor, "They are stealing our sheets." My friend and I ran as fast as we could and went into the barracks before anyone could get down three flights of stairs. That was the end of it. No one complained and nothing else was ever stolen from anyone in my company.

A Big Pussy

One afternoon when everyone returned to the barracks after attending classes all day, the company commander started talking about what we call a big cat. He said a big cat is just a big pussy. As he was talking, I had no clue, what his point was. Then he revealed a greeting card I had received from a girl I had dated. The card said something about being a tiger. I wondered how in the hell he got his hands on that. After embarrassing me, he lectured everyone about locking their lockers with the padlock provided to them. I was tired of always unlocking the combination lock and figured out that I could push the lock shaft to the point just before it latched and it would look as if it was locked when, in fact, it really was not. The company commander walked through the barracks and hit every lock to see if anyone was doing this, and I was caught.

Gun Training

Gun training was part of the curriculum, but not the rifle exercises like in the Army has. There was one afternoon at a firing range when everyone shot about ten rounds of a 22 at a target. That was it for rifle training. It did not matter if we hit the target; they just wanted us to learn how to handle a gun. By the time we started the class, the weather was turning colder, the classroom was warm, and it was hard to stay awake. This was an important class and not one to sleep in. We were taught how the big guns worked and how to load them.

In the heat of combat, a person who is not in ordinance or a gunner's mate may be required to operate the guns to defend the ship.

That happened to a Third Class Cook, Doris Miller, on the USS West Virginia during the Japanese raid on Pearl Harbor. Lieutenant Frederic H. White ordered Miller to help him and Ensign Victor Delano load the unmanned #1 and #2 Browning 50 caliber anti-aircraft machine guns aft of the conning tower. Miller was not familiar with the machine gun, but White and Delano told him what to do. Miller had served both men as a room steward and knew them well. Delano expected Miller to feed ammunition to one gun, but his attention was diverted, and when he looked again, Miller was firing one of the guns. White had loaded ammo into both guns and assigned Miller the starboard gun. Miller fired the gun until he ran out of ammo. He was awarded the Navy Cross for his action.[3]

Testing

Everyone was tested for math, English, and abstract reasoning skills to determine a job type. A couple weeks later, everyone was interviewed for the job he would be doing for the next four years. I was asked if I wanted to be a "penis machinist."

I said, "What is that"?

The chief petty officer said, "Corpsman"! They were attached to the marine's corps who were right in the middle of the action, and were dying in Vietnam at a high rate.

I said, "I was promised electronics."

The chief petty officer said, "Oh, yeah, I guess your scores were high enough for that, so we'll put you in for that."

When I graduated from boot camp, there was a big ceremony with marching and a Navy band. Colors were presented and speeches were given by Navy officers. My family came to see the event. Upon graduation from boot camp, all Seaman Recruits (E1) were promoted to Seaman Apprentice (E2). The next day everyone was sent home on a boot camp leave before going to his next duty station.

Pay

I was paid $88.00 per month while in boot camp. Out of that money, they took enough to pay for that clothing issue I received on day one. I was sent home with about $250.00 for four months' pay. In my civilian job, I would have earned over $2000 in the same time.

Now that I was a Seaman Apprentice (E-2), I received a pay raise to $106.00 per month. I do not think the civilian population realized how little we lived on. I did not have to pay for food or housing, but I did have to pay for uniforms and everything else I needed or wanted.

National Defense Medal

The National Defense Service Medal was instituted in 1953 by President Dwight D. Eisenhower and was awarded to any honorable active duty service during the Korean War from June 27, 1950 – July 27, 1954, and the Vietnam War from January 1, 1961 –August 14, 1974. The ribbon has a broad center stripe of yellow representing high ideals. The red, white, and blue stripes represent the national flag. The red for hardiness and valor, white for purity of purpose and blue for perseverance and justice. No more than one medal is awarded to a single individual but a bronze star denotes additional awards of the medal. The Navy never issued the medal; you had to buy it yourself. Everyone pinned one on his chest after boot camp graduation. I said, it's a part of the "Everybody gets a medal program" just for showing up.

While I was in boot camp, my bother Larry enlisted in the Army under the officer-training program. When I returned home, he was at Army boot camp.

I was given a two weeks leave after boot camp, but that was extended by 6 days because my friend Steve Spratt decided to enlist and he gave me credit for recruiting him.

Boot Camp Graduation

Chapter 3 - Class "A" School

Christmas 1966

After two weeks of boot camp leave, I reported to Great Lakes Naval Center for Class "A" electronics school on December 3, 1966. Two weeks later, the Navy schools shut down for Christmas leave and I went back home for another two weeks.

I did not go straight home. Barbara Hoover and I had kept in touch, and she invited me to Des Moines, Iowa to meet her family. I took the Denver Zephyr train from Chicago to Des Moines to see Barbara. While in Des Moines, I mentioned that my Aunt Marlene's last name was Hoover and that she lived in Des Moines too. It turned out that Barbara and Marlene's husband, Harold, were related. I spent two days and one night with Barbara's family. Barbara and I continued to write, but she eventually found someone else and I never saw her again. Barbara's mom called Marlene, and Marlene gave me a ride to Knoxville, Iowa to visit my grandparents, Walter and Jennie Maddy. From there I took a Trailways bus back to Sheffield, Illinois.

When I arrived home, my dad was angry that I was not standing by the front gate of the Great Lakes Naval Station to be picked up. My dad was going to surprise me, but I did not know anything about that. I was already gone. My dad said he would never do that again, and he did not. He said, while looking for me, he went onto the Navy base and searched what seemed like every building on the base. I asked him whether there was security and had they let him wander freely. He said they let him go anywhere he wanted.

Wayne, Larry, and Dale

For the Christmas holiday, my older brother Wayne was on leave from the Air Force and my older brother, Larry, was on leave from the Army, and I was on leave from the Navy. This was the only time the three of us were home together while in uniform. This was the last time our whole family of fourteen was together. My family has six boys and six girls plus mom and dad.

Cigarettes

I was not a smoker, but my dad, brothers, and sisters were. Every time I made a trip home, I would stop by the PX and purchase cigarettes for my family members. I would often walk out of the PX with six to eight cartons at one time. I could purchase them at the PX for $1.00 per carton and the regular store price was about $3.00 a carton. I never sold them for a profit; I just did this to help my family save money.

Friends

I met Allan Ebell and Craig Howell at the Class "A" school. We were in the same barracks. Allan had a car and drove Craig and I everywhere during those days.

Allan, Craig and Dale

On off duty weekends I invited them to come to Sheffield and meet my family and friends. Craig was from Kentucky and was pleased to have someone take him in like that. Over the next five months, we made six weekend trips to Sheffield. We had to get a special pass to travel that far. It was about 175 miles to Sheffield from the Navy station. Normally, they only allow a sailor to travel 50 miles from the base when on liberty. That would have been to Chicago or Milwaukee, but not Sheffield.

The Diamond Pin

In January, while on liberty, I walked past Tyrrell Jewelry Store in North Chicago, Illinois. A young salesman who was standing by the door, asked me to come in and look at a brooch pin that was designed by a sailor especially for his mother. He told me a story that went with the pin. It was a gold four-leaf clover with engraved lines emanating from the center and a diamond in the middle.

The diamond was to show the sailor's love for his mother. The four-leaf clover was for good luck and that he may return in good health. The lines from the center meant he would be thinking of his mom no matter where in the world he may be.

He then said, "What about your dad; would you like to get him something too?" He had a tie tack pin with the same design for fathers. The price was about $200 for both of them. I told him at my pay level I could not afford such a thing. He said, "No problem, if you will give me $15.00 down, you can pay the balance over the next 21 months." That is what I did. It would be a $9.00 per month payment. He sent a package to my mother with the paper message included in the box. That piece of paper has long since been lost. My dad wore that tie tack everywhere for the next eighteen years. He was wearing it on the day he died. My daughter now has my mother's pin. I learned that whatever you give your parents will someday be returned to you. I am glad I gave them this gift; I never regretted spending that money.

Milwaukee

Is it possible for the actions of one person to turn you against an entire community? The answer is yes. On a weekend in January 1967 three other sailors and I decided to go on liberty in Milwaukee. I had never been to Milwaukee and thought it would be fun to explore a new place.

The only thing I knew about it was Schlitz Beer was made there. Schlitz Beer had a tag line at the end of their TV commercial that said, "Schlitz, the beer that made Milwaukee famous." Schlitz had another tag line that said, "When you are out of Schlitz, you are out of beer." We used to say, "That is because you drank all the good beer first."

On Saturday morning, we took a train from Great Lakes to Milwaukee. When we arrived, we checked into a hotel. We went to a restaurant for lunch and then did some shopping at the Army/Navy surplus store and the Goodwill store. In the evening, we went to a nightclub that was attended by people about our same age. The place was packed. We took our P-coats into a cloakroom and then mingled among the people. When the night was over, our P-coats had been stolen. The temperature that night was below zero degrees. I was pissed. P-coats were becoming a fad among civilians and they were often being stolen. Besides the fact that it was so cold, I had no coat to wear back to Great Lakes. A new one cost me half of my monthly pay. After walking the streets in Milwaukee in the frigid air without a coat, I was so mad that I refused to go to Milwaukee for anything for the rest of my life. I ended up living in Wisconsin and stayed away from Milwaukee for 25 years except for business trips that I could not avoid. Yes, I held resentment towards the people of Milwaukee for 46 years.

Foggy Night

One Friday night we went to Chicago to visit Craig Howell's sister. She was a flight attendant for American Airlines and roomed with three other flight attendants. The other girls were all working and she was there by herself and appreciated the company and getting to see her brother. She was working on a jigsaw puzzle. I loved to put them together, and it helped pass the time. I insisted that we do not leave until it was completed. That turned out to be about 02:00. The guys would have stayed the night there, but we had to be back to the Navy base by 09:00 for a unit inspection. You know, where we dress up in our dress uniforms and then a higher-ranking person nit-picks us apart for stupid things. Well, anyhow, when we left her apartment, there was such a thick fog that Allan could not see the road. At that point, we would have returned to Chicago, but we had that inspection in the morning. We could not miss it or we would be in deep doo doo. Therefore, Craig sat on the hood of the car and guided Allan to keep him from going off the road or from hitting something or somebody. A trip that normally would take about an hour or less had turned into a three-hour trip. We walked into the barracks about 05:00. After two hours of sleep, we started getting ready for the 09:00 inspection. At 08:00, a petty officer informed us, that the inspection was cancelled. We were free to do whatever we wanted at that time. An hour later, with little sleep, my friends and I jumped into Allan's car and headed to Sheffield, Illinois for the weekend.

Drunken Train Ride

One visit home that turned into a fiasco was when I was bringing Jim Elasesser from Buffalo, New York home. I had called the girls from Princeton, Illinois and told them I was coming home and would they meet me at the train station in Bureau Junction. Before the train left Chicago, Jim wanted to go to a liquor store and get something to drink. He asked me what I wanted and I, not being a big drinker of liquor, did not know. Jim suggested premixed cans of bloody Mary. I said, "That's fine." To me it was like drinking an off flavored tomato juice. During the train ride, I drank six of those cans. When the train stopped in Bureau Junction, I took one step off the train and landed on the tracks on my face. Jim had no idea where he was, or where he was going, or who was going to meet us. Luckily, the two girls were there and saw me fall. I was so drunk I could not even get up. Jim managed to get me into their car and the girls knew where my parents lived, so they took us there. When we arrived at the Maddy house in Sheffield, my dad just laughed. Then he introduced himself to Jim, thanked the girls for bringing me home, and helped Jim put me in bed. My dad never said a word to me about what happened that night.

The Stolen Car

During another of my visits home, I brought Craig Howell and Allan Ebell with me. One of the funniest things I ever experienced in my entire life happened on this weekend. My friends and I were all 20 years old; therefore, we were not legal to buy alcohol. Don Sheldon was 21. I did not really know Don very well, and the other guys did not know him at all. His family lived one block south of the Maddy house. The Sheldon family moved to town after I graduated from high school. Don's younger brother, Larry, hung out with my younger brother Paul.

The guys and I decided to buy some beer for the weekend and have Don make the purchase. We took up a collection of money and gave it to Don. The bar was only three and a half blocks away, and Don said that he would need some help carrying it. I assumed they were walking to the bar, so I asked Craig to go with Don to help carry it. As they left the house, I sat by the window and watched them leave. Across the street was a 1954 mint green Chevy. To my surprise, they walked across the street, got into that car, and drove away. The car belonged to Gerdie McKee, the neighbor. At the time, Gerdie was mowing her lawn and did not notice anything. Her grandson, Deems, who was about six years old, was playing in the front yard, watched as Don and Craig drove off. Quickly, Deems ran to his grandmother and told her someone had just taken her car. She ran into the house and called the police. I sat by the window in disbelief of what I just witnessed. Within just a few minutes, Don and Craig returned, but this time they parked her car on the Maddy side of the street. Gerdie was hysterical and screaming at the two of them. About that time I walked out of the house and asked them, "What the hell are you doing?" Craig said he thought it was Don's car and Don said he thought it was Craig's car. When they were leaving, as they walked across the street to the car, Craig asked Don if he was driving and Don replied, "No, you drive." So Craig got into the driver's seat. After getting behind the wheel, Craig looked at Don and said, "Where are the keys?" Now this should have been a clue to Don that this was not Craig's car, but Don just said, "Look in the visor." Craig flipped down the sun visor, the keys fell into his lap, and he started the car and drove away. Neither one of them realized they had just stolen a car. By now the police arrived. Gerdie, Craig, Don, and I were standing by her car. Gerdie started to tell officer Minert that these people had just stolen her car and she wanted them arrested. I explained to officer Minert that my Navy friends and I were home for the weekend, and it was just a misunderstanding. He agreed there was no harm, so we should give Gerdie her keys and everyone go on their way. I said that was fine, but first we needed to get something out of the trunk. We opened the trunk in which there were four cases of Budweiser. Officer Minert never said a word about all of us being under age. Maybe he did not know our ages. He just let us take the beer, and he left. I laughed about that incident for years.

The Sjoquist Family

One night a group of sailors went to the beach on Lake Michigan by Waukegan, Illinois. I met three sisters that night, Linda, Dolly, and Karen. They invited us to their home to meet their parents the next weekend. They lived in Arlington Heights, Illinois. If you traveled a straight line, it was about 20 miles away. Their mother Ruby had fixed a huge turkey dinner with all the fixings for us. The whole family was very nice and lots of fun. They sure made all of us feel welcome. The father, Eldon, was a custom cabinetmaker. He ran his own business. He told me he needed a partner and, if I was interested, I should call him when I returned from the Navy. He told me that if I married one of his daughters, he would give me his business. I decided later that the whole thing was about getting the daughters married. They asked me to keep in touch, and I promised them I would. I wrote to the family for the next three years and continued to send Christmas cards until the 1980's. Ruby sent me a letter at some point and told me Eldon had died. They were such a friendly family; it was a pleasure to keep in touch. I totally lost track of the girls. I often wondered how my life would have turned out if I had pursued any of that.

More Weekends at Home

On another weekend in April, 1967 my dad asked if I could come home and baby sit with my younger brothers and sisters while he and my mother traveled to a company outing. I agreed to do that and then called a girl I knew in Princeton, Illinois and invited her to bring a couple of her friends to meet my friends for a little party. Well, it escalated to more than a little party. Before I got home many people had somehow heard about the little party, and it became a large party. My friend, Lynn VanAcker was home on leave on that weekend, and he came to the party.

The disciplined sailors got up early on Sunday morning and cleaned up the entire house before returning to Great Lakes. There was no evidence of any party from the previous day. My younger brothers and sisters knew, but never really said anything about it.

My Girlfriend

During one of my visits home in April 1967, I met Dee Dee. She was the girl for me. She was a petite, blue eyed, blonde lady. She was the all-natural, girl next-door type girl. I wanted to be with her more than any woman I had ever known. She said she felt the same about me. The timing could not have been worse because I was going to be shipping out in a couple of months. We talked about me leaving and she said she did not mind waiting for me to return. I thought I would be gone about three years. Dee Dee and I wrote letters to each other every week.

Classes

School was divided into three phases. The first phase was about math and electronic components. This was a repeat of the RETS school that I had already completed, so it was easy.

The second phase was about electronic circuits and concepts, which I had already learned at RETS school.

During one of these classes, the instructor had just finished explaining how an amplifier works. Then he turned to me and said, “Do you understand that?”

I said, “Yes.”

He said, “Explain it to me then.”

When I could not explain it, I realized that when you truly understand something, you should be able to explain it to someone else. Repeating something or explaining something you have learned is a great way to test yourself on your understanding of it.

Something the Navy did that impressed me was to give the final test at the beginning of the class. During the course, they taught everything on that test. At the end, you would take the same test, and by comparing the two, a person could see the progress he made. The Navy would use a test as a means of teaching and learning. I used that same training technique years later when I became a sales manager for Agri-King.

The third phase of my training was building a solid-state AM radio transmitter. This was not a part of the previous training I received at RETS school. All of this training was only an entry-level electronics course and there was still a lot to learn.

Shortly before leaving class “A” school, I was promoted from Seaman Apprentice (E2) to Seaman (E3). My pay had now reached $128.00 per month with an additional $10.00 for sea duty.

Next Duty Station

During the final days of class "A" school, they gave me a choice on what I preferred for my next duty. I did not have to think about that very long because I wanted to go to California and to be on a carrier. That is what I submitted, and that is what I received. The Navy tries to give a sailor a choice because he will be happier if he can go where he wants.

They were looking for people for submarines, but I felt that would be too confining and claustrophobic. The Navy does not want to force anyone into that duty. I toured a sub 30 years later and the smell of diesel fuel permeated the entire boat. I could not have survived those conditions. On sub duty, you are required to learn everyone else’s job. It is a matter of survival. If something happens to one of the sailors, somebody had better know how to do his job.

About a week before the last day, I received my orders for my next duty; it was the USS Kearsarge. When others were asking where everyone was going, I said the USS Kearsarge.

They started to laugh and said, "Oh you get the queer barge."

I said, "What is so funny about that duty?"

They said, "That is where all the queers are" (*the words gay and lesbian, were not used at this time*).

Actually that was not true at all; it is just that "queer barge" rhymes with Kearsarge. I had mentally already taken ownership of my new home and was offended that people would say such things about it. I just ignored their comments and went on my way. Actually, I learned later that the USS Kearsarge has quite a history, and a proud one at that. It turns out there were three of them from 1861 to 1946. The first two were battleships, and the third was a carrier.

Class "A" School Graduation

Chapter 4 - USS Kearsarge History

The First Kearsarge 1861-1894
Sloop-of-War USS Kearsarge

Eighty-six years before I was born, at the beginning of the Civil War, the US government started an emergency shipbuilding program. A 1550-ton sloop-of-war was built to stop the Confederates from bringing goods into the South from European countries. One of those ships was named the USS Kearsarge, named after a mountain in New Hampshire. In the 1650's the mountain had been called *Ca-*

rasarga, which is believed to mean, "Notch pointed mountain of pines." Because of her heroic battle in 1864, subsequent ships were later named in honor of this ship.

She was launched on September 11, 1861 in Portsmouth, NH. with Captain Charles Pickering in command. In February 1862, she left for the coast of Spain to blockade Confederate raiders. Captain

John Winslow took command in April 1863 while still in European waters. On June 13, 1864 in neutral waters, just off the coast of France she confronted the CSS Alabama as it left the port of Cherbourg. The Alabama fired on the Kearsarge, but Captain Winslow and his crew were able to retaliate and within an hour, the Alabama was reduced to a sinking wreck. The CSS Alabama's Captain Semmen surrendered her flag and abandoned ship. The Kearsarge rescued the majority of the survivors, and the rest were picked up by a British yacht.

The Kearsarge returned to New York City and the Captain and her crew were honored by the Chamber of Commerce. Seventeen of the crewmen received the Medal of Honor for valor during the battle. They were: Michael Aheam (Paymaster's Steward), John F. Bickford (Captain of the Top), William Bond (Boatswain's Mate), James Haley (Captain of the forecastle), Mark G. Ham (Carpenter's Mate), George H. Harrison (Seaman), John Hayes (Coxswain), James H. Lee (Seaman), Charles Moore (Seaman), Joachim Pease (Seaman), Thomas Perry (Boatswain's Mate), William B. Poole (Quartermaster), Charles A. Read (Coxswain), George E. Read (Seaman), James Saunders (Quartermaster), William Smith (Quartermaster), and Robert Strahan (Captain of the Top).

The Kearsarge then went back to the European waters and the Mediterranean in search of more Confederate ships, returning to Boston in August 1866.

When the Civil war was over, she sailed to the South Pacific along the coast of South America protecting American interests there. Four years later, she went to Japan, China, and the Philippines to do the same thing. During this time, she carried Professor Hall's scientific party from Nagasaki, Japan to Vladivostok, Russia to observe the transit of Venus. For those who do not know, this is when the planet Venus appears as a small, dark disk moving across the face of the Sun. This occurs once every eight years.[2]

The USS Kearsarge met her demise on February 2, 1894. She was wrecked on Roncador Reef near the Solomon Islands in the South Pacific. They tried to rescue her, but the effort proved fruitless and she was stricken from the Navy list later in that year.

The Second Kearsarge 1898-1941

Battleship USS Kearsarge (BB-5)

Forty-nine years before I was born, the second USS Kearsarge (BB-5), a battleship named by an act of Congress to commemorate the famed sloop-of-war, was launched on March 24, 1898 by the Newport News Shipbuilding Company. Traditionally, battleships were being named after states. This was the only exception ever made to honor the outstanding career of the first Kearsarge. It had a displacement of 11,525 tons and a complement of 657 men. She was sponsored by Mrs. Herbert Winslow, wife of Lieutenant Commander Herbert Winslow, USN, the only surviving son of John A. Winslow, commander of the first USS Kearsarge during her famous battle with the CSS Alabama. This battleship was commissioned February 20, 1900, Captain William M. Folger in command.[4]

USS Kearsarge became the flagship of the North Atlantic Fleet cruising down the Atlantic seaboard and in the Caribbean. (*A flagship is a vessel used by the commanding officer of a group of naval ships, flying a distinguishing flag. It is the lead ship in a fleet of vessels, typically the first, largest, fastest, most heavily armed, or best known.*)[5] She sailed into Guantanamo Bay, Cuba where, on December 10, 1903, the United States took formal possession of the Guantanamo Naval Reservation (the same place Iraq terrorists are being held today). From there, she led the North Atlantic Battleship Squadron on goodwill calls to many European ports. In December, 1907, the USS Kearsarge, now commanded by Captain Herbert Winslow became a part of the famous "Great White Fleet" of battleships sent around the world by President Theodore Roosevelt. All of the ships were painted white and sent around the world. The cruise, lasting more than a year, was an overwhelming success showing the US flag and spreading good will. This dramatic gesture impressed the world with the power of the U.S. Navy.[6]

The USS Kearsarge was at the Philadelphia Navy Yard for modernization for several years. On September 17, 1915, she left there to land a detachment of Marines at Vera Cruz, Mexico. She joined the Atlantic Reserve Fleet in February, 1916 and served as a training ship for Massachusetts and Maine Naval Militia until World War I, then trained armed guard crews and naval engineers. During WWI, she spent most of her time searching for enemy submarines.

In May, 1919, she embarked on a Naval Academy Midshipmen training voyage for three months to the West Indies. A year later she was decommissioned for conversion to a crane ship and a new career.[7]

Designated AB-1 August 5, 1920, the USS Kearsarge received an immense revolving crane with a rated lifting capacity of 250 tons and rendered invaluable service for the next twenty years. One of the many accomplishments was the raising of the sunken submarine USS Squalus off the New Hampshire coast.

USS Squalus (SS-192), a diesel-electric submarine built at the Portsmouth Navy Yard, Portsmouth, New Hampshire and commis-

sioned there on March 1, 1939, suffered a catastrophic valve failure during a test dive off the Isle of Shoals at 07:40 on May 23. Partially flooded, the submarine sank to the bottom and came to rest keel down in 240 feet of water. Navy divers and salvage ships responded quickly, and the following day began operations to rescue the surviving 32 crewmembers and one civilian from the forward sections of the boat. At 11:30 on May 24, USS Falcon (ASR-2) lowered the newly developed McCann rescue chamber--a revised version of a diving bell invented by Commander Charles B. Momsen--and, over the next thirteen hours, all 33 survivors were rescued from the stricken submarine. On September 13, after long and difficult salvage operations, Squalus was raised by the USS Kearsarge and towed into the Portsmouth Navy Yard. The boat was formally decommissioned on November 15, renamed Sailfish on February 9, 1940, and recommissioned on May 15, 1940. She went on to receive nine battle stars for WWII and a Presidential Unit Citation on her tenth patrol.[8]

On November 6, 1941 the USS Kearsarge was designated "Crane Ship Number One," giving up her illustrious name which was to be assigned to a mighty aircraft carrier. However, she continued her yeomen service and made many contributions to the American victories of World War II. As "Crane Ship Number One," her name was struck from Navy List June 22, 1955, and later she was sold for scrap.

The Third Kearsarge 1946-1974
Aircraft Carrier USS Kearsarge
(CV-33, CVA-33, and CVS-33)

One year before I was born the USS Kearsarge (CV-33), a 27,100-ton Essex-class aircraft carrier, was built at the New York Navy Yard.[9]

What is an Essex-class ship? In drawing up the preliminary design for Essex class ships, particular attention was directed toward the size of both her flight and her hangar decks. Aircraft design had come a long way from the comparatively light planes used in carriers during the 1930s. Flight decks now required more takeoff space for the heavier aircraft being developed. Most of the first-line carriers of the pre-war years were equipped with flush deck catapults, but owing to the speed and size of these ships, very little catapulting was done except for experimental purposes.

With the advent of war, airplane weights began to go up as armor and armament got heavier; aircrew complements also increased. By the war's end in 1945, catapult launches became more common under these circumstances, with some carrier commanding officers reporting up to 40% of launches by catapult.

The hangar area designers came in for many design conferences between the naval bureaus. The supporting structures to the flight deck were required to carry the increased weight of landing and parked aircraft. In addition, they were to have sufficient strength to support the storing of spare fuselages and parts (50% of each plane type aboard) under the flight deck and still provide adequate working space for the men using the area below.

One innovation in Essex-class was a portside deck-edge elevator in addition to two inboard elevators. The design of the deck-edge elevator was adopted after it proved successful on the USS Wasp (CV-7) experiments. Hauling aircraft by crane up a ramp between the hangar and flight decks proved too slow. The Navy's Bureau of Ships and the Chief Engineer of A.B.C. Elevator Co. designed the engine for the side elevator. It was a standard elevator, 60 by 34 feet in platform surface, which traveled vertically on the port side of the ship. The design was a huge success; it greatly improved flight deck operations.

There would be no large hole in the flight deck when the elevator was in the "down" position, a critical factor if the elevator ever became inoperable during combat operations. Its new position made it easier to continue normal operations on deck, irrespective of the position of the elevator. The elevator also increased the effective deck space when it was in the "up" position by providing additional parking room outside the normal contours of the flight deck, and increased the effective area on the hangar deck by the absence of elevator pits. In addition, its machinery was less complex than the two inboard elevators, requiring about 20% fewer man-hours of maintenance.

Improvements were made in the ventilation system, lighting systems, and the trash burner design and implementation.

Essex carriers had better armor protection, better facilities for handling ammunition, safer and greater fueling capacity, and more effective damage control equipment than predecessors did. They were designed to limit weight and the complexity of construction. For instance, extensive use of flat and straight metal pieces, and of Special Treatment Steel, a nickel-chrome steel alloy that provided the same protective qualities as Class B armor plate, but which was structural rather than deadweight were incorporated.

The original design for the Essex-class assumed a complement of 215 officers and 2,171 enlisted men. However, by the end of World War II, most crews were 50% larger than that, with 300 officers and 3,300 enlisted men.

The tactical employment of U.S. carriers changed as the war progressed. In early operations through 1942, the doctrine was to operate singly or in pairs, joining for the offense and separating when on the defense. The theory was that a separation of carriers under attack not only provided a protective screen for each, but also dispersed the targets and divided the enemy's attack. Combat experience in those early operations did not bear out the theory.

As the new Essex-class carriers became available, tactics changed too. From war experience, the Navy learned while under attack, the combined anti-aircraft fire of a task group's carriers and their screen provided a more effective umbrella of protection against marauding enemy aircraft than was possible when the carriers separated.

When two or more of these task groups supported each other, they constituted a fast carrier task force. Lessons learned from operating the carriers as a single group of six, as two groups of three, and three groups of two, provided the basis for many tactics that later characterized carrier task force operations with the evolution of the fast carrier task force and its successful employment in future operations.[10]

The USS Kearsarge (CV-33) was commissioned in March 1946 and arrived in her homeport, Norfolk, Virginia on April 21, 1946. She spent her first year of service in training operations in the western Atlantic and Caribbean with Captain Francis J. McKenna in command. She departed Norfolk on June 7, 1947 enroute to the United Kingdom on a midshipmen training cruise. She returned to the United States in August of that year.[11]

In 1947, the year I was born, Ensign John W. Lee, the first African American Officer in the regular Navy, was assigned to USS Kearsarge.

For the next ten months, the USS Kearsarge engaged in maneuvers until June, 1948 when she departed Norfolk, Virginia for the Mediterranean. During this tour of duty with the sixth Fleet, the USS Kearsarge was placed on alert to assist in keeping the peace in the Ar-

ab-Israeli area. After her return to the United States on October 2, the USS Kearsarge operated on the East Coast and the Caribbean.

On January 27, 1950 the USS Kearsarge departed for the West Coast and arrived at Puget Sound Navy Yard in Bremerton, Washington on February 23. On June 16, 1950 she was decommissioned there for a modernization overhaul to enable her to handle jet aircraft. The USS Kearsarge was the first carrier to launch jet powered aircraft from her deck.

Recommissioned on February 15, 1952 the USS Kearsarge now had a stronger flight deck, new island and many other changes to her appearance and capabilities. After a shakedown cruise she departed San Diego on August 11, 1952 and headed for Hawaii. Following intensive flight training in Hawaiian waters, the USS Kearsarge sailed for her First Far East Cruise. She arrived in Yokosuka, Japan on September 8, 1952 to begin her first combat assignment.

Any sailor who served on the USS Kearsarge battleship (BB-5) or the carrier (CV-33) from 1898 to 1951 would have thought being in the Navy was a piece of cake and lots of fun, but that all changed in 1952. In all that time, neither of those ships had been in combat.

Korean War

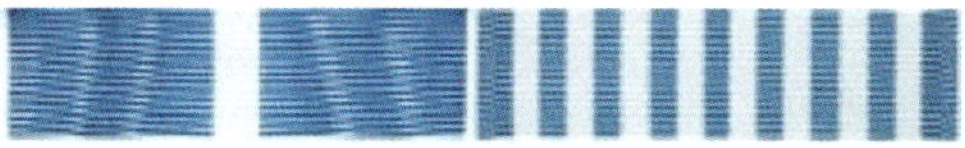

In 1952, when I was sitting on the couch with my mother and learning about war and death, the Kearsarge was going into war with North Korea with Captain Louis B. French in command. She made a Korean War combat cruise from September 14, 1952 – February 22, 1953. During this time, she was reclassified to attack aircraft carrier, and re-designated CVA-33.

F2H Banshee in 1951 over Korea

The USS Kearsarge operated with Task Force 77 off the East Coast of Korea during the Korean War. The mission was interdiction of supplies and transportation facilities. In addition, she was to provide close air support for United Nations troops. Jet aircraft were new to carriers at this time. Naval jets were slow and underpowered. Straight wings hampered performance. The Russian built MIG-15 outclassed the F2H-2 Banshee, and the F9F-2 Panther, which flew from USS Kearsarge.

F9F-2Panther

The AD-4 Skyraider and the F4U Corsair were propeller driven aircraft still in use at this time. Despite the threat from the faster and more nimble MIG-15, Kearsarge air wing carried out their missions. Weather permitting, air operations were carried out daily while on line.

AD-4 Sky raider

The ship and her crew performed very well despite work force shortages, cold winter weather, and long on line periods. Problems with communications,

F4U Corsair

and adjustments to operating with jet aircraft were handled as they arose in a timely manner. A leaking flight deck caused by jet exhaust plagued the ship, but the crew took care of it as best they could until proper repairs could be made in port. Every challenge that arose was handled by a dedicated crew.

Ship's Log[12]

Sep 1952 to Feb 1953

14 Sep 52: CV-33 underway from Yokosuka, Japan to Korean operation area.

16 Sep 52: The Kearsarge joined Task Force 77.

17 Sep 52: An AD-4N Skyraider was hit by ground fire. Shrapnel hit the crewmember who was transferred to the naval hospital, Yokosuka, Japan.

18 Sep 52: A hung rocket broke free from a F2H-2 Banshee during an arrested landing. The rocket skidded across the flight deck and struck two crewmembers that were seriously injured.

30 Sep 52: An F4U Corsair on a strafing and bombing mission was seen to go into a sudden inverted dive following a strafing run. The aircraft crashed into the water off the coast. The pilot was reported killed in action.

4 Oct 52: While on a combat mission, an F4U Corsair was attacked by a MIG-15 near the coast. The aircraft went into a steep dive and crashed into the water. The pilot was reported killed in action.

5 Oct 52: An AD-4N Skyraider crashed into the water immediately after a catapult launch. The pilot and crewmember were picked up by a helicopter and neither suffered any significant injuries.

7 Oct 52: After making a bombing run over Yongpo, North Korea, an AD-4 Skyraider's engine caught fire for unknown reasons. The pilot was forced to ditch the aircraft, but was recovered and returned to the ship with only minor injuries.

11 Oct 52: The ship's helicopter, while being parked on the deck edge elevator with rotors engaged, suddenly tipped over. Four men died, and eight others were injured in this accident.

16 Oct 52: The 33 year old Commanding Officer Bruce Simonds of Attack Squadron VA-702, while flying an AD-4 Skyraider, crashed into the water after an apparently normal deck launch. He was seen floating free of the plane, seemingly alive, with his life jacket inflated. Helicopters and destroyers were on the scene immediately, but were unable to locate him. It was assumed that he was tangled in his parachute and drowned. He was reported missing in action.

18 Oct 52: Departed Task Force 77 proceeded to Yokosuka, Japan.

20 Oct 52: CV-33 arrived in Yokosuka, Japan for rest, recreation, and yard availability. USS Kearsarge was now reclassified CVA-33.

28 Oct 52: CVA-33 underway from Yokosuka, Japan for the Korean operating area.

31 Oct 52: The Kearsarge joined Task Force 77.

1 Nov 52: An F9F-2 Panther was presumed crashed in the water, cause unknown. The pilot was not recovered and was listed as missing in action. An F4U Corsair crashed while on a combat mission. Probable cause was enemy anti-aircraft fire. The pilot was declared missing in action.

8 Nov 52: An F4U Corsair crashed on a combat mission over North Korea, probably due to anti-aircraft fire. The pilot was declared missing in action.

21 Nov 52: An F9F-2 Panther crashed on a combat mission, probably due to anti-aircraft fire. The pilot was declared missing in action.

23 Nov 52: An F9F-2 Panther caught fire while being catapulted. The pilot made a controlled water landing. He was recovered by helicopter with only minor injuries.

4 Dec 52: Detached from Task Force 77 and proceeded to Yokosuka, Japan.

6 Dec 52: The Kearsarge moored in Yokosuka, Japan.

16 Dec 52: The Kearsarge departed Yokosuka, Japan and headed for the Korean operating area.

18 Dec 52: The Kearsarge joined Task Force 77.

22 Dec 52: An AD-4 Skyraider engaged barrier on landing. The aircraft sustained class "C" damage. No personnel were injured.

27 Dec 52: An F4U Corsair from Fighter Squadron VF-884 ditched in the water due to an engine failure. The pilot was not injured. An F9F-2 Panther from Fighter Squadron VF-721 engaged barrier on landing. Aircraft suffered class "C" damage. No personnel were injured.

28 Dec 52: An F9F-2 Panther from Fighter Squadron VF-721 crashed in the water near the formation. The pilot was recovered uninjured. An AD-4 Skyraider from Attack Squadron VA-702 was hit by ground fire while over Kaesong, North Korea. The pilot received minor injuries from shrapnel.

30 Dec 52: While on a combat mission, a F2H-2 Banshee from Strike Squadron VF-11 crashed 15 miles west of Wonsan, Korea. The pilot was declared missing in action.

4 Jan 53: Departed Task force 77, proceeded to Sasebo, Japan.

5 Jan 53: 08:03 Anchored in Sasebo, Japan. 14:59 departed Sasebo, Japan and proceeded to Hong Kong B. C. C.

8 Jan 53: The Kearsarge moored in Hong Kong Harbor.

16 Jan 53: The Kearsarge departed Hong Kong B. C. C. and headed for the Korean operating area.

20 Jan 53: The Kearsarge joined Task Force 77.

22 Jan 53: While making a normal arrested landing, a F9F-2 Panther discharged one round of 20mm ammunition, which fatally wounded a plane director on the flight deck.

23 Jan 53: The 34-year-old Commanding Officer CDR Denny Phillips of Strike Squadron VF-11, crashed his F2H-2 Banshee at 29-16N 127-15E and was declared missing in action.

28 Jan 53: While on a night heckler mission, an AD-4 Skyraider from VC-35 failed to return. The pilot and crewmember were both listed as missing in action. Flying a radar-eyed version of the AD Skyraider, the hecklers have the job of lousing up the enemy's night shift. Swooping out of the darkness and pouncing on Red concentrations

and traffic, they are causing the Commies more headaches and blood-shot eyes than all the Motherland's vodka.

8 Feb 53: While on a mission in North Korea, an AD-4 Skyraider from Attack Squadron VA-145 was shot down by ground fire. The 24 year old pilot, LTJG Donald Hugo Hagge was declared missing in action.

22 Feb 53: Departed Task Force 77 and proceeded to Yokosuka, Japan.

25-28 Feb 53: Up keep at Yokosuka Navy Yard in Yokosuka, Japan.

28 Feb 53: The Kearsarge is now underway for CONUS via Pearl Harbor.

Post Korea

While I was playing in the sand box, the USS Kearsarge made her Second Far East Cruise in 1953. This time she was sent to keep watch on the uneasy truce in Korea. She also kept a watchful eye on the Formosa Straits to keep Mainland China from interfering with the Chinese Nationalists on Taiwan. She returned to The United States on January 18, 1954 and performed training exercises off the California coast.

Her Third Far East deployment started when she departed San Diego on October 7, 1954. Starting on February 6, 1955 and for seven days the USS Kearsarge, operating with the Seventh Fleet, watched over the Nationalist Chinese evacuation of the Tachen Islands and supported the fleet in successfully evacuating 18,000 civilians and 20,000 military personnel from the islands. She returned to San Diego on May 12, 1955.[13]

During the years 1956-57 when I was learning to swim in the pond on the farm, the Kearsarge was again being modernized, receiving an enclosed "hurricane" bow and an angled flight deck to better equip her to operate high-performance aircraft.

Pictured here is the USS Kearsarge in San Francisco bay in 1957.

The USS Kearsarge was assigned a new role in October 1958, becoming an ASW (antisubmarine warfare) support aircraft carrier with the new designation CVS-33. In that capacity, she operated ASW fixed-wing airplanes and helicopters to protect the fleet against the threat of hostile underwater attack. Following intensive training in her new role, the carrier sailed on September 5, 1959 for Seventh Fleet operations on her Fourth Far East Tour. Early in her tour Japan was hit with a violent typhoon, and the USS Kearsarge played an important role in providing relief to the victims. Her planes landed parties of medical and supply units while her crew and air group donated clothing and money to the distressed people. After participating in SEATO exercises and Seventh Fleet operations, she cleared Yokosuka, March 3, 1960 for her homeward voyage.[14]

Three days later in stormy waters 1,200 miles off Wake Island, four Russians were rescued after drifting 49 days in disabled landing craft. They were flown back to their country after the USS Kearsarge arrived at Alameda, California, March 15; the carrier received thanks from the Soviet Union for this gesture.[15]

In October of 1960, USS Kearsarge was sent back to sea only hours after returning to Long Beach after conducting a two-week period of air operations. A Russian submarine had been spotted near the mouth of the Columbia River between Washington and Oregon, and the USS Kearsarge was the only CVS available. She left Pier Echo without the aid of a tugboat. After getting the Air Group aboard, she headed north to meet the sub, low on fuel and supplies, and with most of one liberty section left behind. The submarine was followed to the Bering Sea where a severe storm made it impossible to conduct air operations, and the sub could no longer be tracked. The storm was so bad that waves were actually breaking over the flight deck. The USS Kearsarge returned to Long Beach, California with cracked seams and broken fire mains.[16]

A year of training operations preceded her next deployment from San Diego, which began March 3, 1961. The antisubmarine carrier steamed for the fifth time to the Far East deployment. The destination was Southeast Asian waters as the Communists intensified their effort to overthrow the government in Laos. The power and determination of the Seventh Fleet was observed by the enemy and the crisis eased. Peace won out.

After six months in the Far East, the USS Kearsarge returned to the United States, arriving in Puget Sound in Seattle, Washington on November 1, 1961 to undergo the second phase of her modernization. Following her modernization, the USS Kearsarge performed training exercises off the coast of California.

Krashbarge

From early 1962 to July 1963 while under the command of Captain E. P. Rankin, the USS Kearsarge was involved in three different collisions with ships, a destroyer escort, an oiler, and a passenger liner. The oiler USS Mattaponi (AO-41) and USS Kearsarge were both slightly damaged in a collision during refueling.[17] While none of these collisions was the fault of Captain Rankin, the ship became known jokingly as "Rammin Rankin's Krashbarge" and "The Mighty Kay-Runch." At least it was not being called the "Queer-Barge"

While steaming in dense fog off Long Beach on March 12, 1962, the New Zealand passenger Liner Oriana collided with the USS Kearsarge. The USS Kearsarge received a 25-foot gash about ten feet aft on the starboard side. Oriana suffered a 20-foot hole near the bow and a small fire broke out in her boiler room; the Oriana was repaired at Los Angeles.[17] Based on the

fog experience that I had back in Chicago in 1967, I can really understand how disastrous it can be when it happens on the open sea.

Space Program

About the time Larry crashed Fuzzy's car, the Kearsarge was preparing for another cruise to Hawaii. On August 1, 1962, she left Long Beach to take her place in the space program. She proceeded to the Pacific missile range about 1800 miles northwest of Hawaii.

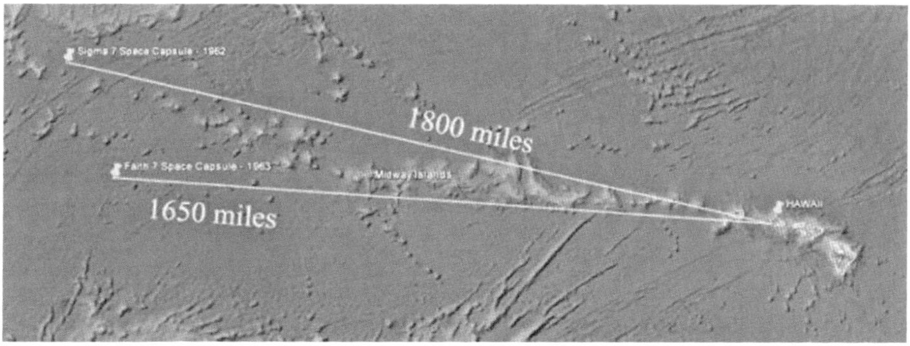

The USS Kearsarge played her role on October 3, 1962 when she retrieved Mercury orbital flight astronaut, Walter Schirra, and his space capsule, "Sigma 7", she went to Hawaii for his flight back home.[18]

The USS Kearsarge operated on training exercises for six months before she entered Pearl Harbor on April 29, 1963 to prepare for another mission for the space program. On May 16, 1963, USS Kearsarge picked up astronaut Gordon Cooper and his capsule "Faith 7" about 1650 miles west of Hawaii after 22 orbits of the Earth. A sonic boom was heard by everyone on the Kearsarge just as Cooper entered earth's atmosphere. Cooper had traveled 600,000 miles, more than enough to go to the moon and back. The capsule landed about five miles directly in front of the Kearsarge.

When he arrived at Pearl Harbor, Commander Cooper dropped a wreath on the USS Arizona Memorial from a helicopter. After a parade in Honolulu, he went to Iolani Palace for a reception with the Governor of Hawaii.[19]

There is a 66-page report on the internet about the recovery; it is called "MA-9 USS Kearsarge Recovery Report." It can be found on Google. The report was confidential, but is now declassified. Everything about the space program was confidential in the 1960's because we were in a race to the moon with Russia. We did not want them to know the details. This report covers minute-by-minute details.

Vietnam War

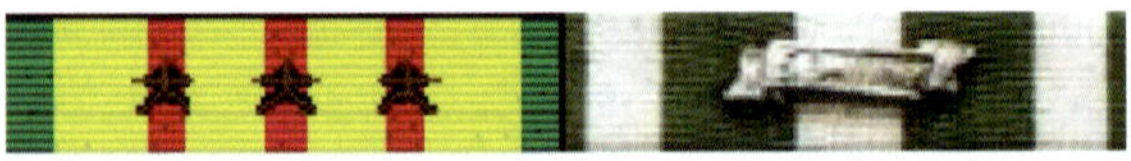

After the Kearsarge returned the space hero to Pearl Harbor, she departed on June 4, 1963 on her Sixth Cruise to the Far East. Operations with the Seventh Fleet included keeping watch on the unsettled problems in Southeast Asia. The USS Kearsarge returned to Long Beach, December 3, 1963 for training exercises off the California coast.

Six months later, June 19, 1964, the anti-submarine carrier was deployed on her Seventh Far East Cruise. Upon arriving in Yokosuka July 30, the USS Kearsarge was dispatched to the South China Sea. This trip was a response to the North Vietnamese patrol boat attack on U.S. destroyers in the Gulf of Tonkin. While U.S. Navy planes destroyed North Vietnam oil and supply depots, the USS Kearsarge provided anti-submarine protection for the Seventh Fleet. The decisiveness of American action persuaded the Communists to delay their objectives for the time being; the Kearsarge returned to Long Beach, California on December 16, 1964.[20]

After being overhauled during the first half of 1965, USS Kearsarge operated on training exercises off California until she departed Long Beach for her Eighth Far East Cruise between April 1965 and December 1965. I was just signing up for the draft at this time.

The Kearsarge made its Ninth Far East Cruise on June 9, 1966. She traveled to Pearl Harbor, and then to Japan. From there it was off to "Yankee Station" where she operated off Vietnam from

August 8 through October 24. There she provided anti-submarine protection for the ships of the "Tonkin Gulf Yacht Club" (as the fleet was known). I was in boot camp at Great Lakes at this time.

The next day she headed for the Kuala Lumpur area and anchored in the Strait of Malacca on October 30, 1966. It was on this trip that the crew of the USS Kearsarge became shellbacks. A sailor who crosses over the equator is called a shellback and those who have not are called a pollywog. A pollywog is a tadpole. When a US Navy ship crosses the equator a time-honored ceremony takes place, the sailors are put through an initiation and given a Shellback Certificate as proof that he has been through the process. King Neptune and his Royal court put the pollywog through an initiation and hazing that he will never forget. It is officially recognized by service record entries indicating date, time, latitude and longitude and sailors count the experience as a highlight of their naval career. I had always hoped we would cross the equator, but I am still a pollywog.

She then returned to Subic Bay, Philippines for liberty. After liberty call, they returned to "Yankee Station" on November 5 and operated there through the 23rd. On November 24, she steamed for Hong Kong. Then it was off to Japan. From there she headed home ending her cruise on December 20, 1966. I will be a part of the crew the next time she sails to Vietnam.

Back in Long Beach

In the spring of 1967 while I was in class A school, the yard period continued. Eventually the ship underwent sea trials. As the ship progressed, USS Kearsarge aided in the training of new A4 Skyhawk pilots in carrier landings. The ship normally did not carry jets, but her hydraulic catapults were capable of handling A4 Skyhawks. When one of these aircraft came in a little too low and the front wheel hit the back of the ship, the landing gear collapsed. This put a small dent in the fantail edge of the flight deck. That dent was still there when I boarded the ship.

In 1967 LT David Hawtin was taking a writing class and asked me to read a story he had written. It was about his experience during that incident. He was standing watch on the bridge and witnessed it. He said that with the front gear broken the plane tumbled off the angled flight deck and the right wing hit the ship. The plane flipped. When the pilot ejected, he was shot into the ocean, breaking his neck and killing him.

Sailors Hurt in Blast

Just prior to my boarding the Kearsarge, this article was in our local newspaper.

FIRE ON U. S. CARRIER

San Diego, June 4, 1967 — the Navy reported today that an explosion and fire rocked the aircraft carrier USS Kearsarge yesterday, injuring three crewmen during maneuvers off the southern California coast.

Reported in satisfactory condition suffering from burns were Torpedo men Kenneth Becker and Bruce Anderson and fireman Charles Hurley, who were taken to San Diego Naval Hospital.

Authorities say the fire aboard the Long Beach based carrier resulted from an explosion in the ship's oxygen-nitrogen compressor department.[21]

Chapter 5 - On The USS Kearsarge

Arriving in Long Beach

It was the first time I had been in Long Beach, California, or anywhere in California for that matter. I was not sure where to go when my plane landed at LAX. I caught an airport shuttle that took me to Long Beach and the driver told me where I could catch a bus to the Navy base. Once at the Navy base, the main gate security guard directed me to the pier Echo where the ship was moored.

As I walked up to the Kearsarge, I stood for a moment and stared at the massive size of this ship. I was going to call this place home for the next few years. I boarded the USS Kearsarge on June 26, 1967. We were taught in boot camp to salute the American flag when we step onto a ship. We are then to salute the quartermaster and ask for permission to come aboard the ship. I have never known him to deny anyone permission to board, so I guess it was just a formality. I was not sure what to expect when I arrived, but the quartermaster greeted me, and guided me to the master at arms office who guided me to the right place and took care of my rack (bed), and locker assignment.

When I first boarded the USS Kearsarge, it was under the command of Captain Benjamin C. Tate. After we departed for our tour to Vietnam, there was a change of command ceremony and Captain Creighton W. Cook became the twenty-second commanding officer of the Kearsarge. They both served in WWII, and were both highly decorated for their service. Creighton Cook had been involved in the Polaris missile design.

During my first couple of months on the ship, I was placed in the aft part of the ship where new people are placed until they determine where you are going to be permanently. At the time, I knew I was going into OE division but there were no vacancies in their berthing area. As soon as a bunk opened, I was moved to be with the other men in my division.

I was put in OE (Operations Electronics) division. The men in my division were all very nice and welcomed me. There were 37 men in my division. Some of them were on the previous cruise in 1966. It did not take long for me to fit right into the group. The men in my division were designated as one of the four: DS - Data Systems Technician, ETR - Electronics Technician Radar, ETS – Electronics Technician Sonar, or ETN - Electronics Technician Communications. I was an ETN. Today ETN designation means Nuclear Trained Electronics Technician. In today's Navy I would be called a CTM - Communications Technician (Maintenance Branch)

The Ship

This ship and its operations were a marvel all its own. After reading about the previous history, it is amazing that the ship functioned as well as it did. The Kearsarge has been ever changing. The ship was built and then refitted at least three different times. The sailors that run the ship were forever changing too. Every year she had a new Captain and a lot of the crew changes almost every cruise. In fact, while I was there, men were transferred on and off the ship even while at sea. OE division had 45 men on the 1966 cruise and only 14 of them were still on board for the 1967 cruise. OE division had 37 men on the 1967 cruise with 23 of them being new. That was a 62% turnover from the previous cruise. When a sailor's service time is up, he wants to go home, so they fly him off the ship even in the middle of a combat zone.

One of the reasons the ship worked so well with all the changes is that Americans are proud people who take ownership of their space. If a person looks at the enormity of the ship's operation, the United States Navy does a remarkable job. There were 2500 ships company, and when the air group is on board, the crew increases to 4000 people. This is four times larger than the little town of Sheffield, Illinois that I came from. It was a floating city with four mess halls, one for the officers, one for the chief petty officers, one for the first and second-class petty officers, and one for everyone else. They operated 24/7, even while in port.

It had a library, a chapel, a laundry, soda fountain, barbershop, bakery, tailor shop, and a clothing store. We even had a movie theater, a closed circuit radio, and TV station. Just like home, it has a fully equipped hospital, a dentist, a police department, a jail, a post office, a general store, an electrical generator plant, sewer plant, water filtration plant, a fire department, plumbers, and electricians.

We also had a print shop that published a daily, "Plan of the Day." It told us which duty section was on watch for the day. It listed the OOD (officer on deck), uniform to be worn for the day (blue work clothes, white or dress blues), ship's travel status, ship's daily routine, ship's crew birthdays, announcements and notices, and the food menu for the day. They also published a 16-page monthly news magazine called "*Kearsaga*." Each issue featured a different female model. Sharon Tate, who was murdered by the Charles Manson clan in 1968, was featured in the spring 1968 edition. The magazine had many female photos. It had stories about places the ship had just visited and positive happenings on the ship.

On the military side, the ship was an airport, an airplane hangar, a tanker, and an armory, with ship fitters, offices, and computers. Hundreds of bombs, rockets, torpedoes and shells were stored in her many magazines.

In the ship's stores, there were over 60,000 supply items worth over seven million dollars. It included everything from jeweler's scissors to airplane engines.

I walked onto the flight deck to look around. It was 894 feet in length. That is about three football fields long. It was 214 feet wide. There were two hydraulic catapults for launching airplanes and five arresting cables for stopping incoming aircraft. Those cables looked like little threads from the sky, but up close, they were the size of your wrist. An airplane coming in at 120 mile per hour would catch a cable and sixty yards later roll to a stop. The Kearsarge could travel at a top speed of more than 30 knots. One knot = 1.15077945 miles per hour, so the Kearsarge was capable of traveling faster than 35 MPH. The ship would turn into the wind when launching and retrieving aircraft. If the head wind was 20 MPH and we were traveling 30 MPH; the wind across the flight deck would be 50 MPH. An airplane making an approach at 120 MPH would be doing only 70 MPH relative to the ship.

There were three airplane elevators. One was located on the aft starboard side. Another was on the mid port side just at the end of the angled landing strip. The third was located at the front of the ship between the two catapults.

After all the modernization overhauls, she became a 31,000-ton displacement. She had a draft of 30 feet. There were four screws (props) powered by eight boilers generating 150,000 horsepower.[1]

There was a lot of activity going on when I first boarded. Things were being repaired and new things were being installed. Many new people like me were just getting familiar with the ship. For the next two weeks, I spent most of my time just being acclimated to the ship and my surroundings. Then we went to sea for ten days on what is called a shakedown cruise to test everything in every department on the ship.

The ship had some of the most sophisticated electronics found anywhere in the world. There were three radio rooms on the ship, with as many as 20 radios in each. Some were transmitters and some were receivers. There were many backup units in the radio systems. The radio rooms were in three different locations so that in an attack we would have backup communications. There was lots of radar equipment also. There was radar for scanning the air, one for scanning the ocean surface, and another for checking aircraft approach speed. Then we had tacan, which was used for navigation. There was IFF (Identification Friend or Foe), facsimile, sonar, and computers. Most of the electronics were either in the super structure or in the first deck under the flight deck. The electronic parts supply room was also on the same level. Radio transmission rooms and radio equipment rooms were not the same. Radar rooms and radar repeater rooms were not the same. One of my first general quarters assignments was in the CIC (Combat Information Center), where the radar repeaters were. There was a lot of activity going on there. I watched on the big board as they plotted enemy planes coming in and as we launched our aircraft to retaliate. It was an interesting place to be. I was required to have a secret clearance to be in these rooms.

My more permanent job became the parts supply man for the OE division. This job was almost the same as my civilian parts job at John Deere. Instead of farm equipment, I was dealing with electronic parts.

Because of where I was located on the ship, I was always near the flight deck and the super structure. This was great because I saw daylight almost every day. Some of the men who worked below the hangar deck never saw the sun except in port.

My berthing area and the mess hall were below the hangar deck, so I was always walking between the super structure (above the flight deck) and below deck, which took me through the hangar deck.

I wanted to photograph my life in the Navy for memories of places I went and things I witnessed. Having access to the flight deck allowed me to photograph airplanes in action. I photographed almost every activity I could without getting into trouble for doing it.

Required Movie Films

Shortly after boarding the Kearsarge, I was required to watch a film about venereal disease. It was enough to make you never want to have sex with anyone. I have since learned that during the Civil War more men were lost due to venereal disease than the war itself. During WWII, prostitution was a huge business in Honolulu and venereal disease was a big problem then. The US government did not outlaw prostitution; they just required the girls to be checked every week to keep disease under control. This film was not a new one; it was made for the WWII sailors.

We were required also to watch a propaganda movie about the reason we are going to fight in Vietnam. This was a Hollywood made film. They wanted each of us to feel sorry for the South Vietnamese people. I do not know if any of it was true. It was a great propaganda film to make us feel there was a purpose for our being involved in the war. Why would our government show this film to us and not the American public? I had not seen it before this. My family back home would have felt better about my being there if they had a clear understanding on why we were doing this.

The third film required concerned what to do if captured. I had seen this film while in boot camp. If you have an urge to run, you should do it immediately while you still have strength. Once you are captured, the enemy will wear you down by depriving you of sleep and food and you will never have the same strength again. If you have a weapon and you are not strong enough to use it, you should give it to someone who can use it.

Over the next few months, we also watched a movie called "The Third Challenge" about the cause of the gorilla war and our purpose for being there. I feel it was just propaganda and it meant no more to me than that.

We watched a victory at sea movie about WWII in the Philippines. It was interesting. We were also required to attend a safety lecture occasionally. I remember attending a propaganda flic on the USSR and a lecture on facsimile equipment.

A couple of days before arriving in Subic Bay, we watched a TV lecture about liberty in Olongapo. It was quite interesting; it was about counterfeit money, narcotics, classified information, and benny boys. An interesting fact that was brought out is that five million dollars a month is spent there.

On another day, we had a lecture on radar repeaters and a repeat on venereal disease. We had a lecture on burns and another victory at sea movie.

We watched a movie on survival at sea. It was good information worth knowing.

Before entering Sasebo, we had a lecture on liberty in Sasebo. It was interesting, but he did not tell us anything I did not already know. I attended a very interesting hour training lecture on US Navy history.

Shots

Shortly after boarding the Kearsarge, I was told to report to the forecastle to get shots. Shots were mandatory. I told them I did not want the shots and was refusing to take them. They just simply replied that that is okay, but you cannot get off the ship if you do not take them. Oh, crap! Seeing the world was one of the things I wanted to do while in the Navy. I guess I had better comply. I was given a shot for Typhus, Polio, Flu, Smallpox, Typhoid, Cholera, Tetanus, and Yellow Fever.

Chief I. M. Sheethead

There was one of those bully chief petty officers in my division. Chief I. M. Sheethead (not his real name) came to me during my first time at sea and started physically pushing me around. I had not said anything to provoke him; he just started doing it for no reason. He was like the schoolyard bully. Maybe he was marking his territory, or just trying to be tough. I do not know what his point was, but I knew I did not like him. I told some of the other guys about the incident; and they told me he did that to everyone. There were two witnesses to this incident: they wanted to smack the guy but knew better. No one liked him. I think he thought that because he had a higher rank than everyone did, he could do this type of thing.

One of the things I despised about the military was the chain of command thing. You are taught to respect anyone with a higher rank than yourself. If they give you an order, you are supposed to comply. That is fine. I understand that and do not have a problem with it. What I do have a problem with is abuse of that privilege. Higher rank does not mean you have a right to be mean or abusive to others. You can still show them respect. Only if a lower rank refuses to help with the task, should any other behavior be used. The Navy has rules to deal with people who refuse to carry out an order. Chief I. M. Sheethead was not giving me any order to carry out. He was just being a bully.

We decided that he had an inferiority complex and he took it out on us by trying to make us feel inferior to him. He did this by always disagreeing with us. If we said black, he said white and vice versa. One time while I was cleaning a storage locker, he made the comment, "Boy this place is a mess," implying that somehow it was my fault. I did not think that comment was necessary. Sure, it is a mess when you have everything lying out on the deck as you are sorting through stuff. I just ignored him and continued with my work. He once made a comment that if I would get busy and work we would not have so much bad equipment. He implied that it was all my fault the equipment was failing. When he would enter the room, I would leave. One afternoon I spent the entire day in the ship's library just so I would not be near him. When he was around, he would cut everyone in the room as much as he could just to be irritating. I could not stand him. Finally, he went home on an emergency leave, and I do not remember seeing him again after that. My productivity increased after he was gone.

Making Coffee

The new kid on the block gets the crap job that no one else wants. That happened to me when I was the newest guy in OE division. I was told that first thing every morning I was to make a pot of coffee. We had a 20-cup coffee maker. That did not seem so difficult. All I had to do was run down the passageway, get the water, put the coffee in the basket, and turn it on. How hard could that be? I took the coffee maker to the water and dumped the old grounds in the trash. When I looked in the pot, it was black with crud from months of making coffee. I do not think anyone had cleaned that pot in years. Remembering what my dad had said about doing the job a little better than expected, I grabbed a scouring pad and scrubbed until it was shining like a new one. The next morning the Second Class Petty Officer wanted to know who made the coffee. I proudly stood up and said, "I did."

He looked at me and said, "You ruined our pot," "Do you know how long it took us to get that pot broke in? Take him off that detail." That was the end of my coffee making job. My dad was right: if you do your job better than expected, you are promoted. I made only one pot of coffee in three years on the Kearsarge.

My Hawaiian Friend

On July 19, we were back in port for four weeks before leaving for the Far East.

During those weeks in port, we were given liberty two nights out of three. While in Long Beach one evening, I was minding my own business, walking down the sidewalk. There was a man walking in the same direction as me, but I was not paying any attention to him or anyone else. As I walked, a girl came running out of a building and ran right between the two of us. She pushed the other man and me to the side as she ran.

He looked at me and said, "Looks like she's in a hurry."

I said, "Yes, it looks that way."

He asked me if I was in the Navy. I told him I was. He then asked me where I was from. I told him Illinois. He told me he had lived in Chicago. I told him my sister lived there. He asked if she was married. I told him she was dating a doctor. He wanted to know the doctor's name. I thought to myself, what difference it would make if he knew, he would not know him anyway. He said, I am a doctor and I was in Chicago for my internship. When I told him the name, he said that he knew him.

He looked like maybe he was Japanese. I asked him where he was from, and he said Hawaii. His name was Bob Maehara. My first concern was he was gay and looking for a gay partner and I was reluctant to befriend him. He told me he was not gay and he never made a pass toward me. Bob and I stopped in a bar and continued to visit for a couple of hours. Bob was a native Hawaiian and a great ambassador for the state of Hawaii. He asked what my schedule was for upcoming cruises. When I told him we were leaving for Hawaii and Vietnam in a few weeks, he told me he would be home with his mother about that time. If I would call him when I arrived, he would show me around the island. On August 17, 1967, the Kearsarge started her Tenth Trip to the Far East.

Pearl Harbor

When the ship arrived at Pearl Harbor, a tugboat slowly took us into the harbor. It was in the late afternoon, and the sky was a golden color. The sailors stood at parade rest and lined the edge of the flight deck. It was an impressive sight to see. During the first stop at Pearl, I was a part of that, but on the second stop I was not included and was able to photograph it from the superstructure. Once the ship was secured and the gangway was in place, we were free to go on liberty if we were not on watch.

Off the port side about seven hundred yards away, was the USS Arizona Memorial. I did not really know Pearl Harbor history at this point in my life, so I did not have much interest in taking a closer look at the memorial. I took a picture from the flight deck for my collection.

Don Ho

When we arrived in Pearl, I went to the Navy base nightclub with my friends and did not bother to call Bob Maehara. The next day I called Bob and he picked me up within an hour. He told me I should have called the night before because he was at a pool party at entertainer Don Ho's home and many girls were swimming nude in the pool. I guess I should have called. That was a window of opportunity which had opened, closed, and never opened again.

Bob introduced me to his mother, brother, and sister. He drove me on a tour around the island of Oahu. It included surfing at Waikiki beach, the sugar cane fields, Pali Overlook, Chinaman Hat Island, Dole Pineapple Pavilion, swimming at Sunset Beach on the North shore, a picnic at Wiamea Falls Park, and a night with Don Ho. Bob and Don Ho were high school friends.

Bob took a couple of my friends and I to Duke Kahanamoku's nightclub. It was located in the International Market Place Square where Don Ho was entertaining. When we arrived, there was a line of about 75 people ahead of us. Bob told me to wait at the end of the line while he went to look things over. A few minutes later when he returned, he told me to follow him to the front of the line. Bob told me I did not need any money because everything was on him. He knew the maître d', so when we entered we were able to go straight to a table in the front. The maître d' told Bob that he did not have to pay for anything, that it was all on him. In the end even the maître d' did not have to pay. During the show, a multi-millionaire sang with Don and announced that he was buying all the drinks. What a night; I did not pay, Bob did not pay, and the maître d' did not pay. Bob took me to the last show of the night because he knew Don would get wild and crazy during that show, and he did. After each song, Don would lift his glass and say, "Suck-Em-Up." They sold many drinks because of that. Don sang his signature song, "Tiny Bubbles."

I took Bob on board the Kearsarge and gave him a tour of the carrier. I took him into every nook and cranny that I was allowed to go. He ate supper with me in the mess hall. He had never been on a US ship before, so I am sure it was a thrill for him also.

Bob invited me out to dinner at John Dominis Restaurant in Honolulu. It had a perfect view of Diamond Head and Waikiki Beach. It was the place to go for seafood in Hawaii. It was my first time to eat lobster; Bob insisted on it. I went back to John Dominis' in 2009 and it still looked the same. In 2012 the structure was torn down. It was replaced with a new more modern look. When you are in Honolulu, it is definitely a place you should go to eat seafood. If you are not hungry, at least go there for a cocktail to see the view.

Chapter 6 - Life at Sea

Peeking Over the Flight Deck

During my first trip to sea, I was curious about what was happening on the flight deck. Therefore, I went out on the catwalk and looked over the flight deck. There were no planes taking off or landing at the time, so I thought it would not be a problem. A flight deck officer saw me and yelled, “Get off the catwalk.” I went back inside, but he followed me in and lectured me for doing it. He ordered me to come back on the flight deck with him. He told me to pick up two of the wheel chocks and then run a lap around the flight deck. Two of those chocks weighed about 50 pounds. Running 500 hundred yards with fifty pounds in your arms is enough to make you remember you do not want to do this again.

A month or two later on this cruise, I was in the supply room when something hit the flight deck hard. Something had fallen off an airplane when it was landing. We learned later that a missile had come off the plane at touch down and skidded across the flight deck into the ocean. The place where it skidded off was about where I was standing when I peeked over the flight deck. There was a reason for not letting anyone stand on the catwalk while planes were landing and taking off. On September 18, 1952 while the Kearsarge was in Korea, that same thing happened.

Colored Shirts

The crew wore a variety of brightly colored shirts on the Flight Deck. Who were they, and what did they do? All of the Yellow Shirts were Flight Deck Officers and Plane Directors. These were the only people on the flight deck authorized to move aircraft or to give hand signals regarding taxiing the airplanes, or moving them. Blue Shirts were the flight deck crew who always worked under the direction of a Yellow Shirt Director as part of a designated crew. They were plane handlers; they chocked planes, pushed planes, scrubed the decks, and did all the tough jobs on the flight and hangar decks. Green Shirts were Catapult and Arresting Gear people. They operated, and maintained the catapults and arresting gear. Squadron Aircraft Maintenance people also was green. The Brown Shirts were Squadron personnel, in particular Plane Captains (no ship's company wore brown shirts). Red Shirts were the Aviation Ordinance crews. They loaded and “armed” those weapons on the aircraft. Sometimes they were referred to as “BB Stackers,” or “Ordies.” The Purple shirts were Aviation fuels people, referred to as “Grapes.” They fueled aircraft and, operated and maintained carrier avgas, jet fuel, and the aviation lube oil system. The Checkered Shirts were Squadron maintenance Quality Control people. They were responsible for final overall checks of airplane readiness before launch. The White shirts were Mail and People Handlers for carrier on board delivery (COD). The White shirts with a Red Cross were Medical Corpsmen assigned to flight deck during flight quarters. They were part of the crash crew.[1] The men who operated the catapults were called shooters. They shot the planes off the flight deck.

Dear John Letter

This happened a few months into the cruise. While at sea in the Gulf of Tonkin, a message came over the PA: “Man over board.” When this happens, everyone is to go to General Quarters and reports in so they can figure out who is missing. Immediately a helicopter is launched for rescue. The ship is stopped and the rescue team searches for the missing person. The search goes on for 48 hours before they give up. In this case, we were there for 48 hours and the man was never found. A few days later, I was dumping garbage off the fantail. Someone is always on watch at the fantail 24 hours a day. The man on watch at that particular moment was the same one who was on watch the day the man jumped overboard. He told me a seaman walked up to the rail with a letter in his hand. He said the man dropped the letter on the deck and jumped over the rail. He said he reported the event immediately and then picked up the letter and looked in the ocean behind the ship to see if he could see the man. He said he could not see him. He thought the turbulence from the ship’s screws pulled the man under. He thought the man had drowned right away. He looked at the letter; it was from a girl back home who was dumping him for someone else. He apparently was so distraught over it that he no longer wanted to live. I said right then that no woman is that important that I would even consider such a thing. What a way to die in a combat zone. It had nothing to do with the war.

According to DC3, Sonny Richardson, the man who jumped, hit the master at arms before jumping. Sonny said he also reported "man over board" from his workstation.

As I mentioned before, I was emotionally involved with Dee Dee. She and I were keeping in touch. We were smitten with each other. She and I wrote letters to each other every week after I arrived on the USS Kearsarge. She would send photos of her and kept me up to date on her life and I did the same for her. While in Hawaii, I purchased a pink coral pendant for her. It was something I could afford and it was unique to the area I was visiting. It was something special for someone special in my life. About a month after the man overboard/dear John incident, Dee Dee sent me a Dear John letter. This happened in March 1968, almost one year after we met. She told me she had met someone else and she did not want to wait for me to return. There was no way I was going to jump over board. I was disappointed and heart broken, but committing suicide - Nah! I felt she must not have really felt the way she had described or she would not be doing this. While at sea, I was not able do anything about it. If she did not care anymore than that, I was better off without her. As the saying goes, "the sea is full of fish."

Turning in

At about 20:00 the boatswain's whistle would sound over the ship's 1MC (public address system). Then there would be an announcement to dump trash and garbage over the fantail.

At 21:00 hours the boatswain whistles would sound again and another announcement, "Lights out; all hands turn in; maintain silence about the decks; the smoking lamps are out in all berthing spaces."

"Now here is the Chaplin with the evening prayer." "Let us pray, Oh god in child, for more secure in our people and all generations, let us have peace, comfort, and rest, never be cold, and never forget continued loving concerns for our every need, our God, Oh Lord, Amen." This may not be word for word because the sonar noise in the background made it hard to hear.

Most nights the sonar would be turned on and we would hear it all night long. It would beep, and then a few seconds later it would beep again. The beeps were rather loud. Sometimes the patterns would change and there would be an echo. Sometimes it would be a couple of high pitch whistle sounds and then a couple of low pitch whistles. After time you would get used to it and it would lull you to sleep.

Airplane Mishap

One evening in 1968 at just about dusk in the Gulf of Tonkin, one of my co-workers told me an airplane had just fallen off the flight deck. I quickly grabbed my camera and went onto the flight deck to take some pictures of the incident. One of the flight deck crew was towing an airplane from one side of the deck to another. When he backed up, the front of the plane went off the flight deck and dropped into the gun tub below. By the time I arrived on the scene, the ship's crane was connected to the plane and started to lift it up. I was running around on the flight deck as if I was some official photographer taking pictures from different angles. The flight deck officer happened to see me and asked if I was the ship's official photographer. When I told him no, he told me to get the hell off the flight deck. He probably thought I would send my photos home and who knew where they could end up, possibly the front page of the local newspaper. That would have been embarrassing for the Navy. I did learn from this that if you look official, you could do things that otherwise would have been questioned by someone. The picture did not end up in the paper, but it did end up here in my book.

Mysterious Metal Box

One day I returned to my office and found, right in the doorway, a large metal box with handles on both sides. It was about the size of a two-foot cube. It was very heavy; I could barely lift it by myself. I had no idea where it came from or who put it there. It was in the middle of the doorway and I had to step over it to get into the office. That was very annoying. My friend Ron Smith from S Division would often come to visit me at my office, and he commented about the need to step over the box when coming in. He asked me what it was and where it had come from. I told him I did not know.

He helped me move it a couple of feet so we did not have to step over it any longer. He said we should throw it overboard. I said that if no one claimed it in a month I would go along with that idea. About a month later, it was still sitting in the same place. Ron said, "Let's throw it overboard." It was dark outside and there were no planes flying, so we could just take it out to the catwalk and let it go. I told Ron it would really make a loud sound when it smacked the water because we were 50 feet above the water line. We decided to rig a rope to the handle so we could lower it to the water. We would have another rope rigged so we could pull the tie to release it when it was close to the water. We tested our theory first and the test worked perfectly. Therefore, we tied the ropes on the handle then carried the unit to the catwalk. It was a job just getting it to the catwalk. We tied our end of the rope to the catwalk and lifted the box over the rail.

I said, "Are you ready."

He said, "Yeah."

So, we let it go. The idea was to let it down slowly, but the handle was not strong enough to hold its weight and it broke off and the unit crashed to the sea. The impact on the water surface was deafening like that of an exploding bomb. The first thought I had was the fantail watch would think we were being attacked and report it to the bridge. I had not thought about the boys in sonar. They had to see an object that big, going down the side of the ship. They could have thought it was a minefield. We grabbed our rope and hightailed back to the office. We waited, thinking someone would soon be asking questions. I thought this could have triggered a general quarters, but nothing happened. I learned later that the boys in sonar had their equipment turned off that night.

About three weeks later LT. Heggie asked me if I had seen his de-humidifier. I told him, no, I had not. He then said it had been sitting right here in the doorway for a couple of months. I thought to myself, "Is that what that thing was?" I had already told him I had not seen it because I did not know to what he was referring. When I learned that it was his de-humidifier, I was not going to change my story. That would have been a court martial for sure. It had been over two months since it was first brought into my office. He never mentioned anything to me about it, nor did he say what was to be done with it. He just shrugged his shoulders and let it go. It was never brought up again. Honestly, I do not think it worked and that was why it was there in the first place. Maybe he was going to get it repaired or maybe he was going to trash it and we saved him the trouble.

General Quarters

Every sailor was assigned a specific battle station. For some it was right where he normally worked. For others it was a place not manned under normal circumstances. I worked two different GQ stations during my time in the Navy. Initially I worked in ESO (electronics supply office) and later I was transferred to CIC (combat information center). I liked the CIC station best because that is where the radar repeaters were and there was a lot more action. One time during GQ, things were lax; I played with the repeaters trying to learn all I could about their operation. All air traffic was monitored there. I watched as enemy airplanes came in and our planes met them. When we took out the enemy, the officers would cheer just the same as when we scored a touchdown in high school. That struck me as a strange reaction to having just killed someone. In the movie "Tora! Tora! Tora!" the Japanese sailors did the same thing when the airplanes returned from their strike on Pearl Harbor. Muslims did the same thing when the World Trade Center was hit on 9/11, and Americans did it when Osama Bin Laden was killed.

On shakedown cruises off California, we usually knew when there was going to be a GQ drill. When the GQ bell rang, the boatswain would announce over the 1MC, "All hands man your battle stations." Once you arrived at your battle station, you were to report in over sound powered phones that you were secure. It was important that you go to your battle station immediately because all hatches were closed, and we were not allowed to open them. It was impossible to move about the ship at that time. The captain would start a stopwatch when the bell sounded and stop it after every person on the ship was secure at his battle station. It usually took ten to fifteen minutes. If you were in the chow line two decks below the hangar deck and your battle station was two decks above the flight deck, you had better run or you would not make it. Often when we knew in advance, people would already be at their station before the bell rang. This really gave the captain a false time. One time when I was running through the hangar deck, I saw others just walking. I was annoyed that they did not take this more seriously.

One day while on Yankee Station the GQ alarm rang out, and we were not told in advance of a drill. It was for real! The men were running in every direction and were serious. The captain announced we were all secure in eight minutes, a record time. When the men thought it was for real, they got serious. After that, I always said that when America is under attack, everyone takes it seriously, and it will fight back to the best of its ability. The 9/11 attack on the World Trade Center proved that to be correct. Volunteers signed up by the hundreds right after that happened. America is a peaceful nation, but if attacked, we will fight back vigorously to win - or until death.

We went to GQ on average about four times per month. Generally, GQ would last for only one or two hours. Most of that time I would wear sound powered headphones and about every fifteen minutes, I would report that I was still at my station and that everything was secure. Other than that, I would do nothing during GQ. If there was a fire anywhere on the ship, we went to GQ. Sometimes the GQ was a rehearsal for missile attacks. In all my years in the Navy, only once did we have GQ during lunchtime; we were served C-rations.

One time the GQ alarm rang at 00:30 when I was sleeping. I got up in the dark and grabbed for my shirt and pants that hung next to my rack only to find they were not there. The man in the rack below me had put on my clothes by mistake. He was shorter than I was, so the pants were extra-long. When he realized they must not be his, he returned them. After getting my clothes back, I slipped into my trousers, shirt, and shoes and buttoned them up as I ran to my GQ station. I tied my shoe after I arrived at my station. It turned out there was a fire on the flight deck. After the fire was out and everything secured, we were released. I went back to bed until reveille at 07:00.

At Sea Operations

There is always a lot of activity on an aircraft carrier. For the most part, airplanes are either taking off or they are landing. This takes place nearly 24 hours a day when on Yankee Station. While I was on the Kearsarge, there were times that other things were going on, like UNREP (underway replenishment). Two different operations took place. One was taking on ship's stores, and the other was transferring fuel. These processes were similar. It started by stringing a cable between the two ships. This was done by using a shotgun to shoot a string over to the other ship. A rope tied to the string would be pulled back to us. Then a cable would be tied to the rope and pulled back to them. Then we had a cable between the two ships. On that cable, we could pull a hose across for fueling or a pulley for stores transfers. It was an interesting process to watch. Once when transferring stores, they lost a pallet of goods over the side due to poor working of the high lines. Sometimes they would not use a cable because goods were transferred by helicopter. On average, we had an UNREP about every third day.

Sometimes we were taking on oil or gas from a tanker ship and sometimes were supplying oil to a destroyer. The ships were powered by a thick black crude oil. When we took on aviation gasoline, things got a little more dangerous because that stuff is highly flammable. The smoking lamp was out during oil and gas transfers. We did not want to blow ourselves up. With airplanes flying around the clock, we used a lot of aviation gasoline. The Captain just charged it to his American Express Card and then gave the bill to the American taxpayer. (Laughing Out Loud, - that is a joke.)

When ships stores were brought on board, usually to the flight deck, most of the work was manual. Food arrived on the flight deck on pallets full of 50-pound boxes, and was hand transferred to the lower deck storage compartments. All the boxes were labeled, so when something that was readily edible, such as pears or peaches came through the line, the guys would stash a can somewhere to eat later. The cans were gallon size, so there would be enough to share among the whole division.

One time while anchored in Long Beach, frozen steaks were being loaded and hundreds of pounds were stolen. No one person took much, but many took a little and it added up to a lot. No one was ever caught for snatching stuff because no one was going to snitch on his friends. Looking back on this, there must have been millions of dollars of goods stolen from the Navy stores. The tradition of snatching edible food has been passed from crew to crew. With the large turnover of personnel, no one ever did it more than once or twice.

While we were in Long Beach thousands of dollars of copper was stolen from the ship in the middle of the day. I do not know if anyone was ever caught for that. It was offloaded to the pier by a crane operator, loaded into a Navy truck, and was never seen again.

Man Overboard

When passing through the hangar deck on my way to the mess hall one afternoon, I saw a marine putting a sailor through the paces by making him do jumping jacks, push ups, etc. The sailor was in the brigg for some misbehavior. About 20 minutes later on my return to my office they were still at it. As I walked by them, the sailor got up and ran over to the side of the ship and jumped off the port elevator. In a few minutes the ship's whistle blew six times. It meant man overboard. It was announced that someone went over on the port side. The ship was moving about 10 to15 knots at the time. In just a few seconds we were quite a distance from him. The waves were three to four foot swells, and it was really hard to see anyone. I watched from the catwalk and saw two smoke flares but saw no one. The ship immediately launched a helicopter and about that time, someone yelled out, "There he is!" I looked out across the ocean and he was just a small speck. When a person is in the water, he could easily be overlooked by the search team as they look very small. I surely never would want to go overboard. The helicopter spotted him almost instantly and rescued him. A few days later, the man was given a dishonorable discharge from the Navy and sent home.

One of the men in my division had been stationed on a destroyer prior to being on the Kearsarge. He told me he had been washed overboard. He was on the deck working on an antenna, when a wave came over the ship and washed him into the sea. He said he had a life jacket on and was safe. They were part of a carrier task force and he was rescued by a helicoptor within a few minutes. He said the waves were five to six foot. Even though he was rescued, he said it was still a scary experience.

Navy Food

I always thought my mother was a great cook. We never missed a meal or starved. She cooked a small variety of foods. Most of her meals were what she had learned from her mother. There were a lot of Swedish dishes. Mom would make fresh cinnamon rolls and fresh baked bread. She would make cakes and pies, too. Mom cooked pot roast, pork chops, hamburgers, hot dogs, chicken, turkey, and fish sticks. We never had prime cuts of meat. I never had lobster either.

I heard sailors complain that Navy food was bad, but most of the time it was okay. There were a lot of dishes that I never had until I went into the Navy. After being in the Navy I had eaten a prime rib, filet mignon, rib eye and new york strip steak, either in restaurants or in the mess hall. I remember different times we were served lobster, shrirmp, or crab legs. When the ship was in port, we would sometimes have the finer steaks, but most of the time it was hamburger and ground beef dishes. The Navy chefs prepared more variety of food than mom ever did. Their menus usually covered the five basic food groups.

For breakfast there were almost always bacon, sausage, and eggs with toast. Sometimes we had SOS (short for shit on a shingle or chipped beef and gravy on toast). There was always orange juice, apple juice, cranberry juice, and milk.

When in port we would have fresh eggs and milk. After being at sea for about ten days the fresh stuff would run out and then we would have powered eggs, milk and potatoes. The Navy food was never equal to a five star restaurant, but it was as good as anyone could expect under the circumstances. It was better than the guys in the Army were getting. One of the complaints was we were getting the worst cuts of meat that were full of grissel.

Another thing about the food that turned people off was cockroaches. One time we had a Hungarian goolash and the guy sitting next to me saw what looked like a strand of something sticking out of it. When he pulled at it, it was a cockroach antenna. That was enough to turn us all off from eating anymore.

For five months I kept a journal on the food. I rated it from very good, okay, bad, and won't eat it. There were only one or two meals in a month that I rated as very good. About 60% of the meals I rated as okay, and the rest were rated as bad or I didn't eat. There were times when the food didn't even look appetizing, and I would skip eating. I once skipped five meals in a row. There were also times that I ended up with diarrhea after eating.

ETN2 Barry Rittle told me a story about being in the chow line when he spotted a cockroach crawling across a custard with a meringue topping. He said the cook just took his spatula and pushed the roach into the custard and covered it up with more meringue. Barry said that for about two months, all he ate were peanut butter sandwiches he made himself to avoid Navy food.

ETN2 Ron Brown brought his own cans of Beanie Weenies on board and stashed them in his locker in case the food was not good at times. Then he could eat Beanie Weenies. Ron reminded me about the lime green Kool Aid that sailors called bug juice. He said he hasn't drunk Kool Aid since. Come to think of it, I haven't either.

Water Hours

When the ship left the harbor, the water tanks were full of fresh water. Water was used for drinking, cooking and bathing for 4000 men. They also used fresh water to power wash the airplanes. When the ship was on station, the fresh water supply would get low. Evaporators were used to make fresh water from sea water. The old boilers on the ship couldn't keep up with the water demand. One time we went on water hours because of a leak in the water system. To save water the captain would declare water hours. The showers and sinks would be turned off and on only the drinking fountains would be left on.

When we were on water hours, I could take a shower at only designated times. We were to get wet, then turn off the water. Then lather up with soap and scrub down. Then turn the water back on to rinse off. The whole process was limited to three minutes per person. The master at arms would stand guard at the showers and time us with a stop watch. If we went over the three minutes we were put on report. I don't know of anyone who was ever put on report. A shower that ran longer than three minutes was called a Hollywood shower. I had never heard that term before. I understood what was going on, but I did not like it. I remember times when we were not able to shower for three days in a row. While in the South China Sea with 95 degree weather, we would be in need of a shower by that time. After about 15 to 20 days at sea, we ended up on water hours during our remaining time at sea. Once the ship was back in port, we could go back to taking a Hollywood showers again.

Believe me, every shower I have taken since I left the Navy has been a Hollywood shower.

Our sister ship, the USS Bennington (CVS-20), had her sailors taking salt water showers. They were issued a special soap for use with salt water. Anyone who has ever swum in the ocean knows that you have a salt residue on your body after being in that water.

A sister ship is a ship of the same class as, or of virtually identical design to, another ship. Such vessels share a near-identical hull and superstructure layout, similar displacement, and roughly comparable features and equipment. All of the Essex class carriers were considered to be sister ships.

Mail Call

Mail was our only connection to the outside world; we didn't have access to outside radio and TV. Making a phone call home while in port was cost prohibitive. Mail was very important to us. We didn't have a mail call everyday because of location or logistics and aircraft's ability to fly in or out. One time there was no mail because somone had put it on the wrong ariplane and it ended up on the wrong carrier. I always looked forward to getting something in the mail, even if it was a only a magazine.

I wrote letters to quite a few people. I have already mentioned Dee Dee; I think we wrote each other every week. Dr Maehara's sister Carol sent me a letter almost every month and included a box of chocolate covered macadamia nuts from Hawaii. She was always very nice to me. Ruby Sjoquist wrote me often. My mom and dad wrote at least once a month. Of my family members, the person who wrote a lot was my sister Becky. She sent me a letter nearly every other week. She also sent me a copy of *Playboy* magazine every month. Sometimes when it arrived, it was in rough shape. I think the mail room passed it around and everyone looked at it before they passed it on to me. We would tape the playmate of the month on the bulkhead. The pictures in those days were not as revealing as they are today. I liked it better that way; it left something for the imagination. When I received Christmas boxes with cookies, I shared them around the office. That was a big hit with everyone.

ETN2 Ron Brown was temporaily transferred to a destroyer to repair a radio. When he was finished with his work and ready to return, they held him hostage. They would not return him to the Kearsarge until their mail was delivered.

Sometimes mail would be held at a port where we would be going. When the ship would arrive, there would be a stack of mail waiting on the pier. I took a picture of that on more than one occasion. The pile of mail would be five feet high and ten to twenty feet long. There was always a lot of it.

When we were on Yankee Station we could mail a letter for free. All we were required to do was write the word free where the stamp would normally be. My sister complained that the Chicago post office would charge her postage at her end. They were not supposed to do that.

Hazards

Aside from the fact that we were in a war and subject to being attacked, there were many hazards on a carrier. Men who worked on the flight deck were paid an extra $35.00 per month hazardous duty pay. One day a man walked into the prop of an airplane. I believe he died from his injury. I can see how that could happen. A person who works on the flight deck is hearing the engines roar all day long and would soon get used to it and not realize how close he was to a prop.

FN Rodney Shipman was killed while trying to weld some supports to a bunch of steel that was leaning against the bulkhead. He pulled a four foot by eight foot sheet of steel toward himself and leaned them on his chest. When the ship rolled, the sheets of steel fell on him and crushed him to death.

My friend Steve Spratt worked on the flight deck of the USS Enterprise (CVN-65) as an ordinance man. He had two incidents that could have been fatal. He was blown off the flight deck from jet blast. He was forunate that he landed in the cat walk. He did sustain a broken leg from that, but it could have been worse. Another time a bomb dropped onto the flight deck from the wing of a airplane when it landed. Steve ran to the bomb and rolled it off the deck into the ocean. That took a lot of courage because that bomb could have exploded.

My friend Lynn VanAcker didn't make it through the war unscathed either. He was on a ship on the rivers in Vietnam and he was exposed to agent orange when they sprayed the banks of the rivers. He will be living with the effects of that for the rest of his life.

One day a helicopter was taking off and as it started to lift, it flipped to the side and fell into the ocean. The problem was the portside tie down chains had not been disconnected. I believe a couple of people died in that accident. After that incident the man removing the tie down was required to stand in front of the helicopter, holding the chains up so the pilot could see them.

The biggest concern on a carrier is fires. With all the fuel on board that is being pumped into airplanes, there is huge concern all the time. Then there are the bombs and missiles that are being loaded all day long. All the disasters the Navy has experienced over the years has caused it to put into place as many safety precautions as possible. During a seven month cruise, the USS Kearsarge experienced six different fires. Most were extinguished quickly without extensive damage or personal injury.

While I was still in boot camp in October 1966, the USS Oriskany (CVA-34) had a fire erupt on the starboard side of the ship's forward hangar bay. It raced through five decks, killing 44 men. Many who lost their lives were veteran combat pilots who had flown raids over Vietnam a few hours earlier. A magnesium parachute flare exploded in the forward flare locker of Hangar Bay 1 beneath the carrier's flight deck. *Future US Senator John McCain flew from this ship.* I was on a work detail on the USS Kearsarge later in 1967 when we were handling those same flares. We were told to throw it overboard if one of them accidentally ignited.

On July 29, 1967 while I was in Long Beach preparing for my first cruise, the USS Forrestal (CVA-59) was devastated by a fire and series of chain-reaction explosions that killed 134 sailors and injured 161 while operating in the Gulf of Tonkin. An unusual electrical anomaly discharged a Zuni rocket on the flight deck.

The Navy designed a special fill nozzle for the gas hose used when fueling airplanes. It was made so there would not be any fuel spills. Fuel spills would make the ship very vunerable to fires.

In the mid 1970's there was a bad fuel spill fire in the pit area at an Indy 500 race. The next year all of the cars had been fitted with the same fill nozzle used by the Navy aircraft.

Those of us working in electronics were not allowd to wear our dog tags around our neck like everyone else in the military. If we were to lean over to look into electronic equipment and our dog tag fell out of our shirt, we could be electrocuted. Instead, we put them on a key chain and clipped them on our belt. Just a good safety precaution.

Our ship was never attacked by the enemy while I was on board, but there was evidence of our being in the war. A helicopter had gone into Vietnam to rescue a pilot who had been shot down, and when it returned to the Kearsarge, there were bullet holes in it. I do not remember that anyone was injured, but they were shot at.

Another time I remember that we went to GQ when two MIG 17 jets were headed in our direction from Ha Noi, North Vietnam, about 400 miles to the north. Our task force launched jets to meet them. At about 200 miles north of us, the MIG's turned around and went back to their base. For an airplane that can fly mach one, 200 miles is only 20 minutes away. That is why getting to your battle station in ten minutes or less is crucial.

Some of the men on the Kearsarge were awarded Navy Crosses, Silver Stars, Distinguished Flying Crosses, Purple Hearts, Navy Commendation Medals, and Letters of Commendation. On the 1966 cruise over 350 well deserved awards were given out.[2]

My Job

People have asked me, "What did you do in the Navy?" That answer would be different for every sailor. A cook would say that he prepared meals and after the meal was served, he cleaned up and started the next meal. My friend Ron Smith, who worked in the ship's laundry, would say he washed clothes all day, one load after the other. He eventually worked in dispersing. People in the mailroom did the same things a person in your local post office does all day. A yeomen shuffled paper all day.

In the electronics division, we primarily waited for equipment to malfunction so we could repair it. There was some preventative maintenance, but very little of that. So, for the most part, we did nothing. Every technician was assigned to a radio or radar room and was responsible for keeping his equipment running. Everyone was also responsible for housekeeping in his own space. If we had trash to dispose of, we had to wait for an announcement to dump trash over the fantail.

How did we spend our time then? We played cards, checkers, and chess. We read books and magazines, and we studied for our next promotion test. The ship had a library with just about every magazine printed. I would go there to study my lessons for the next rank. Some men had music collections and listened to their favorite artists on their stereo systems they purchased in Japan. I spent a lot of my off time taking pictures.

My job was a parts supply person for the OE division. I kept a log book on the money spent each day. We were given a new dollar allotment each month, and as we spent the money, I would minus from that allotment giving a daily balance. There was a magnatron tube that was used in one of our radar units that cost several thousand dollars each. We would need a new one almost every month. All items were ordered through my office. I would go to the ships store to get them. I was the "market basket delivery man" because I delivered directly to the person who ordered the item. The reason I say item instead of parts, is because sometimes they ordered tools, pens, paper and etc.

When we were at sea, my office was manned twenty-four hours a day. Wayne Erickson and I each worked a 12 hour shift. We would take turns on the night and day shifts. For one week I would work from 08:00 to 20:00 and he would work from 20:00 to 08:00. At the end of the week we would work two 18 hour days to flip from night to day.

Once, when in Subic Bay, we had thousands of dollars left in our budget. LT Heggie told me to go to the base warehouse and get whatever I wanted and use up the money. He told me if we didn't spend all the money, the next month we would be shorted on our allotment. This is the single biggest way the government wastes tax payer money. I spent about five thousand dollars on all kinds of things we didn't even need. He didn't care what I bought, he just wanted me to spend it. I purchased ten Simpson volt/ohm meters that day, and lots of Snap-On tool kits. I purchased a sheet of plexiglass to put on my desk so I could put pictures under it. I am sure that really helped the war effort! If every division on every ship in the Navy is doing this, just think of the waste it totals to every month.

Additional duties were when we would stand watch when in port. Other times we were assigned to the supply department to assist in the loading of ship stores. Occasionally we were assigned to shore partrol duty.

Payday

Once the ship left Hawaiian waters, on the way to Japan, we were paid with Military Payment Certificates or, as they were called, MPC. This was done to keep American green backs out of the black market.

MPC money could be spent only on the ship and on the Navy base. People in other countries would not take it because they could not cash it in at their banks. America would not honor MPC from foreigners.

When we were in a foreign port, we needed to exchange MPC for the country's own money. When we returned to the base, we could exchange any left over foreign money back to MPC.

As the ship approached the USA on the return trip, we could exchange any MPC back to green backs.

In Japan the exchange rate was 360 yen to one US dollar.

The exchange rate for Hong Kong dollars was six of theirs to one of ours.

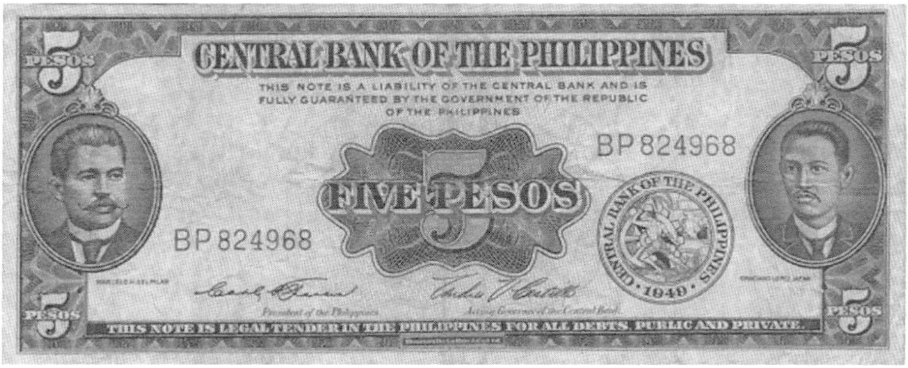

In the Philippines the exchange was four pesos to one US dollar.

Chapter 7 - Port Of Yokosuka

Entering Yokosuka

As the ship entered Yokosuka, Mount Fujiyama could be seen off at a distance on the port side of the ship. Mount Fuji, located on Honshu Island, is the highest mountain in Japan at 12,388 feet. It is

an active stratovolcano, which last erupted in 1707–08. Mount Fuji is about 60 miles southwest of Tokyo, and can be seen from there on a clear day. Mount Fuji's exceptionally symmetrical cone, which is snow-capped several months a year, is a well-known symbol of Japan. It is frequently depicted in art and photographs, as well as visited by sightseers and climbers. It is one of Japan's "Three Holy Mountains" along with Mount Tate and Mount Haku; it is a Special Place of Scenic Beauty, a Historic Site, and has been submitted for future inscription on the World Heritage List as a Cultural Site.[1]

Tokyo Tour

The Kearsarge entered Yokosuka Japan on October 8, 1967, for nine days of repairs and resupply. Each sailor was to stay on board for one out of three days. The other two days we were free to go on liberty call wherever we wanted within 50 miles of the port.

The ship had organized a tour to Tokyo. Tokyo, the place I learned about in the sand box fourteen years earlier. “Bombs over Tokyo.” Yes, 25 years earlier the United States, in April 1942, under the leadership of Lieutenant Colonel James H. Doolittle led 16 B-25 bomber aircraft, each with a five-man crew, to attack Tokyo.[2] I had to see this place so I could learn all about it and its people. I never imagined when I was in the sand box, that one day I would actually walk the streets of this city. What an opportunity for me.

The tour took us to the 1964 Summer Olympic National Gymnasium that was designed with a futuristic look. Today it still looks futuristic. We visited the Tokyo radio/TV tower. The Tokyo Tower was the world's tallest self-supporting steel tower at 1,095 feet. It was definitely among one of the greatest landmarks in Japan. It looked like the Eifel Tower in Paris.

We had lunch at an original Japanese restaurant. We were asked to remove our shoes upon entering. I remember thinking that was going to be a problem because we all wore the same shoe style. With a busload of sailors, how would I find mine when we were leaving? The tables were only about 15 inches high and we sat on a pillow on the floor. I was served beef teriyaki with raw egg sauce. I had never had it before and thought it was very good. We were given chopsticks to eat with, but for those who could not master them, there were forks available.

After lunch, we attended Kokusai Theater for the "Autumn Dance Revue." It was a Japanese version of New York's Broadway. All of the roles were played by women, even the male roles. They sang and danced to music from around the world. The performance was very good.

The American Flag

Visiting in the port of Yokosuka was my first time in another country. I was a foreigner and I felt like a foreigner. I felt like an outsider. I felt like an intruder of another person's country. It is not that I was not welcome because the Japanese people were very friendly to the American sailors. They liked that we spent lots of money when in their cities. The business people spoke English, but most of their people did not know English and I knew only a few Japanese words. We were told by the ship's captain, "You are all American ambassadors so respect the Japanese and their customs." Our uniform did not entitle us to special privileges; rather it obligated us to set higher standards of conduct and performance of duty.

After a day of exploring in the city and dealing with strangers, it was refreshing to return to the ship where I knew people. When I approached the ship, I saw the American flag flying on the fantail. The feeling I had at that moment was something I never thought I would have. As I looked at the American flag, I felt a sense of peace. I felt safety and security. I felt belonging, and I felt as if I was home.

Since that time, I have seen in movies where a person in a foreign country is seeking the American embassy for safety. The movie director shows the American flag. The reason he does this is it gives the viewer a sense of safety. While watching the movie a person can get the same feeling I had.

As a Boy Scout, I was taught how to display and fold a flag. I was taught to treat the flag with respect and never let it touch the ground. If a flag would ever touch the ground, it should be burned and replaced with a new flag. When I returned home from my service in the Navy, I watched on TV as protesters burned and stomped on our flag. I was outraged by this behavior. The Supreme Court ruled that it was okay for them to do this because of their First Amendment right to freedom of speech. In my opinion, you should have the right to freedom of speech only if you have respect for the flag of the country that gives you that right.

The flag flies high at every government building in America, in our schools, at our businesses, on our ships, and on our airplanes. It is on the uniforms of our policemen and our firefighters. It is displayed on NFL players' helmets, and it is in our front yards, and on the caskets of the people who served this great country. The flag is printed on the sides of charitable aid boxes sent to other countries. What feeling do you think those people have when they see those boxes arriving?

I do not care what the Supreme Court ruled; I think they were wrong on this one. If you are on American soil and you disrespect the flag, you are showing disrespect to all patriotic Americans. That should be a crime punishable by time in prison. Americans do not stomp or burn the flags of other countries, not even our enemies.

The American flag represents peace to the pre-school child on a playground. It represents honor to an old war veteran in a nursing home. It represents safety and security to the person being rescued from a disaster. It represents comradery to a group of soldiers and sailors. It represents the Constitution in the halls of congress or in a courtroom. It represents freedom to a political prisoner in another country. It represents America in the White House. It represents all these things at a Fourth of July parade in small town America. How dare anyone disrespect our flag while standing on our soil!

I have a flag that flew over the USS Arizona in Pearl Harbor. I have a flag that flew over the United States Capitol in Washington. I have a flag that flew in Iraq, and I have a flag that draped the casket of a fallen soldier. All of these flags are properly folded and stored in a triangle display box. They deserve respect.

I fly the American flag at my home 24/7 during the spring, summer, and fall months. The ice, snow, and wind take its toll on the flag in the winter months, so I take it down during that time. I have done this since 1973. I did not need 9/11 to happen to fly the flag. When 9/11 happened, Americans flew the flag like never before, so I know that most of them feel the same as I do. They just don't show it every day.

New Camera

One of my first goals when arriving in Japan was to buy a new camera. I had been using a Kodak Brownie Instamatic for the past five years. It took pictures in the three inch by three-inch format. I wanted something with more capabilities than just snapping the shutter. I went to a camera shop in Yokohama, Japan. They had every camera brand and model available anywhere in the world. They had all the accessories also. The big deal about buying in Japan was the price. I found quickly that I could dicker with them on the price. They were already half the price of buying in America, but if you were willing to haggle, they would come down even more. It was funny: the salesman would say, "You numba one, me give you special price." "You numba one, you special customer." That was what I heard at every retail store, restaurant, and bar. He was more willing to throw into the deal extra accessories than to lower the price of the camera.

After I made the deal, the salesman would pick up his abacus and total the purchase. He could give me a total faster than I could do it in my head. The hand held calculator had not been invented yet. I took his word on the total price because not only did I need to add up the purchases, but I also had to pay him in Yen. Figuring out how much I was actually paying was done by dividing the Yen total by 360. I bought about $400 worth of camera equipment for about $185.00 US. I stopped at a money exchange to convert dollars to yen. A $185.00 was about 66,600 Yen. That would be sixty-seven 1000 Yen notes. My wallet was bulging with yen.

I had purchased a Canon 35mm SLR camera with a telephoto lens, special filters, and a flash attachment. I used mostly Kodak Ektachrome film. Ektachrome produced slides with pictures favoring blues and greens. Kodachrome favored yellow and reds. I rarely used Kodachrome.

With this camera, I was able to take pictures in low light and stop action such as stopping the rotorblades of the helo when using high-speed film. I took at least 7000 pictures with that camera before it started to fail. It lasted me until 1992. This was one of my best purchases while in Japan. Many of the pictures I took with this camera have helped keep the memories alive for this book. I am sure that without them I would have forgotten many things long ago.

The Submarine

I was returning from a day in town of taking pictures. I was walking by the dry dock area on the base in Yokosuka. I happened to see that a submarine was in the dry dock. I thought it would be cool to have a picture of a sub in dry dock. I quickly ran over to the sub, aimed my camera in that direction, and snapped a couple of pictures. I then started to walk away when I heard someone yell at me. I turned around and there stood an officer. I do not know where he came from. I did not see anyone just a few seconds earlier. He asked if I had just taken a picture of that sub. I told him no. He said, "Yes, you did - I just saw you." He demanded that I open my camera and expose the film. I complied. I just lost all the pictures I had taken in town. I also lost the sub pictures. I don't know what the big deal was. In 1951, Warner Brothers made a movie called Operation Pacific, starring John Wayne. They showed a submarine in dry dock. I was not going to send the picture to the Soviet Union; it was just for my own collection. That sub was probably scrapped years ago.

Flight Deck

When we were in Yokosuka, Japanese workers were performing repairs on the flight deck. It was at this time that I learned the flight deck was made of wood. I am not sure of the exact dimensions, but it appeared that timbers of about five inch high by five inch wide by ten feet long were bolted to the steel deck. Then they were painted deck gray and coated with a gritty antiskid paint surface.

I thought it was ironic that during WWII, the Japanese were bombing our flight decks and now they were repairing them for us.

Hotsi Bath

Prior to entering the port of Yokosuka, the old timers were telling me about things I should do when in port. The one they insisted on was getting a Hotsi Bath. Of course, I said, "What's that?" They told me it was the greatest massage I would ever get in my life. They told me that I might want to have a few beers before I go. Roy Moruss and I took a taxi to Yokohama. Yokohama was a few miles up the road from Yokosuka and before Tokyo. Roy had made the trip before, so he became my leader. He learned that the word hayaku meant faster in Japanese, so he was telling the taxi driver "hayaku," after we started moving down the road. When he said that, the driver started going faster. So, I said it again and he went even faster. We thought it was fun, so we kept saying "hayaku" and the driver kept going faster and faster. I wondered if they had a speed limit; it did not seem like it. He was going around corners at a speed that threw us all over the back seat. The more he did that, the more we laughed, and the more we laughed the more he did it. This driver was practically going around corners on two wheels. It was a riot.

First, we went to a local bar and drank a few beers. After we were feeling loose, Roy said, "Let's go to the Hotsi bath." We hailed a taxi and he drove us to a hotel. We were greeted at the door by an older woman who offered us a cold beer. We were required to remove our shoes when we entered (a Japanese custom). I was given a basket like the one you get at the public swimming pool and told to enter the door in front of me and strip naked, placing my clothes in the basket. They gave me a couple of minutes to undress. The room I was in was about four feet by four feet. Opposite the door I entered was an exit door. There was a knock on the door and a young Japanese girl opened it. She was a petite girl about four feet five inches tall and dressed in a white outfit like a nurse. She took my basket of clothes and led me by the hand into the next room. She told me to watch my step because the floor was wet. The room was like a brightly lit high school locker room, with benches and, white tile walls and floor. She took me to a whole body steamer. I sat in it as she closed the door. She wanted to know how hot I could stand it. I told her to set it to whatever was normal for most men. I sat in the steamer for about five or ten minutes until I was sweating. She then told me to sit on the stool in the middle of the room over a floor drain. She then started to scrub my back the same way my grandmother did when I was a little kid. She washed every part of my body with soapy water. When she finished the wash, she took a gallon bucket of warm water and rinsed away all the soap. Then I stood up and she towel dried my body, finishing up by wrapping me in a large warmed towel.

I had never been pampered like this before. I just let her do her thing. She grabbed my basket of clothes and told me to follow her again to another room. We went up a spiral staircase into a large room, about the size of a basketball court. On the floor were wall-to-wall mats like the wrestling mats from high school. On those mats were about 20 men who had arrived before me. Each man was at some stage of a full body massage. She first had me lie on my back as she massaged my feet, arms, hands, and face. Then she had me lie on my stomach and she massaged my legs, arms, and back. She stood up and walked on my back, which really felt great. When she finished, she told me I could get dressed whenever I was ready. The bath and massage lasted an hour and a half. I did not even want to get up, let alone get dressed. The answer to your next thought is, “No.” Get your mind out of the gutter. When we left the hotel, we taxied back to the ship, went straight to bed, and slept the night away. I wanted to take that young lady home with me.

I have since had many massages, but none has ever compared to this. I have had them on a cruise ship, at the Venetian Hotel in Las Vegas, on the beach in Hawaii, at health spas, and in my home. The Hotsi bath wins, easily. The Hotsi Bath cost me about $20 and most of these others were a $100 or more. I now knew why the guys were so insistent that I try this. If you are ever in Yokohama, this should be on your list of things to do.

Chapter 8 - Port Of Hong Kong

Rust vs. Silver

When the ship was entering the Hong Kong harbor, many of us were standing by the railing. The water was deep blue and clear. You could see 10 to 20 feet down. Locals were in small boats begging for money. Sailors were throwing coins into the water, and the local boys were diving in and retrieving the coins. It was in January and the water was cold. I do not know how they could do it without getting hypothermia. One of the guys standing next to me picked up a chunk of rust from the deck and flung it into the water. The Chinese boy dove into the water and retrieved it. He got back into the boat and started all over. That piece of rust did not discourage him. That was a good lesson for me: when you are cold and tired and all you get are chunks of rust, you must keep diving until you get a piece of silver.

Hong Kong

Hong Kong is an island off the coast of China. It was a British Crown Colony. The British had a 99-year lease on the island. The lease was going to expire in 1997. At the time that seemed like a long, time away. However, that day has come and gone and the island is now in the control of China. Many feared that China would bring the city down, but they did not. It is as thriving as ever. In Hong Kong you could buy anything, from any place in the world. There were many contraband items such as ivory for sale. You could exchange money from anywhere in the world as well. If you wanted to take the chance, you could buy Yen in Japan and then sell it in Hong Kong and make money on the deal. Money fluctuated, so if your timing was wrong, you could lose money doing that.

Custom Made Uniform

Hong Kong was the place to get a custom made uniform. Many sailors were wearing gabardine uniforms. A gabardine uniform looked sharper than a Navy issued uniform. It was made with a zipper under the arm so it was easier to put on and remove. It was custom measured, so it was a snug fit. The inside was lined with satin.

If we wanted it, they would sew a dragon into the inside of the wrist. Men who had been on the previous cruise were wearing them, so it was easy to know who had been here before. When we returned to pick up our uniforms, the tailor bought supper for all of us. I paid $30 for this new custom tailored uniform. I wore that uniform less than 50 times over the next 24 months.

Kowloon Tour

The Kearsarge arranged a tour of Kowloon. Hong Kong is an island and Kowloon is the mainland China. I took the tour that began with a shuttle boat taking us to the pier in Kowloon.

I remember the tour guide pointing out what looked like an ordinary apartment building. She said that each apartment had two rooms and that ten people lived in each apartment. She said the people sleep in shifts. I thought yeah, that is what you get when you have a billion people in your country. My high school history teacher told me there are so many people in China that you could line them up four abreast and have them walk by you and the line would never end. By the time the first billion people would get past you, so much time would have past there would be new generations of people. It would go on forever. I like the open spaces of America. I am not a New York, Chicago, or L.A. kind of person. I like the lyrics to the Bing Crosby song, "Don't Fence Me In."[1] Everywhere we went on this tour was overcrowded.

We stopped at a forbidden walled city. People were born, raised, and died within the walls. It was built from bricks. The streets/corridors were about four feet wide. The walls were about ten feet high. No pictures were allowed. We could walk into the first section, look around and then we had to leave. It looked like a prison to me. I took a picture outside after the tour, but I did not take any on the inside.

We also stopped by the bay where fishing boats were tied by the hundreds. The tour guide told us these people lived on these boats 365 days a year. She said they ate fish and never left the boat. The same story was, they were born, lived, and died on the boats. Again, this would not be for me. There are people here in the U.S. that live similarly.

Everywhere we went the people wanted money if you took their picture. One little boy, about four years old spotted me taking his picture and he started demanding money from me. If I had a dime for every time someone took my picture, I would have a lot of money today. It is just as bad in Las Vegas; people stand on the street waiting to have their picture taken and then they want a tip for posing for you.

Chapter 9 - Port Of Subic Bay

Olongapo City

We arrived in Subic Bay, Philippines on November 13, 1967. The day before we arrived one of the guys asked me if I wanted to have Chief I. M. Sheethead assassinated. He said if I did, this was the place to do it. He said you could get anything you wanted here and that included having someone exterminated. He said that for $20 I could have someone kill him. I asked if he was serious, and he replied with a "Yes." He told me this had been discussed the last time the ship was in this port. He told me I would not have to concern myself with this because Chief I. M. Sheethead would never go into town. To pull this off you would have to get him there. I never gave this idea a second thought.

I knew the guys were right, though, because the night we arrived in port, the gates were closed and no one was allowed in town. Someone had been shot the night before we arrived. We spent the evening at Cubi Point, an on-base nightclub. On the way back to the ship, we heard gunshots coming from town.

The next day the gates were reopened and we went to town. Immediately outside the gate was a bridge that crossed over the Kalklan river. In the river were young boys and girls standing in flat bottom boats, begging for money. There was a girl who was dressed in a white outfit. She looked to be about 16 years old. One of the sailors picked up a huge rock and dropped it off the bridge. The rock went right through the bottom of the boat and she sank into the muddy water. I do not know why some people are so mean. That sure was not a good image for America.

We were told to guard our wallets and not to wear our ribbons into town. Locals were known to steal sailor's wallets and to rip their ribbons off their uniforms. I thought I was in Dodge City, Kansas when I walked the streets of Olongapo City. The streets were dirt, the sidewalks were wooden, and the saloons had swinging doors, just as the ones I had seen on the TV show Gunsmoke. As I walked down the street, a man flew out of those swinging doors and landed on his back in the street. Wow, this was the Old West all over again. There was a man standing by the man who just landed on the street, and he was wearing a six-shooter in a holster. I was expecting Matt Dillon to show up any second. Maybe this is why the sailor dropped the rock through the bottom of that boat. I had never seen anything like this before.

My friends and I went into one of the bars and had a round of drinks. The local girls came to our table and sat with us. When the photographer came by the table to take a picture, men and women who were not sitting with us gathered to be in the photo. The girls were hustlers and wanted us to buy them drinks.

Wallace, Jack White, Me, & Ron Smith

I saw women doing things in that bar that were like nothing I had ever seen or even thought about before. This was sin city if there ever was one. We had a few more rounds and then went back to the base. I never went back into that town again. Once was enough to last me a lifetime.

Pagsanjan Falls Tour

The Navy arranged a two-day tour to Pagsanjan Falls. It was about 120 miles from Subic Bay. The trip took us through Manila. The falls are located southeast of Manila. We did not stop in Manila other than for a potty break. On the trip we drove through Olongapo City and I took pictures of homes that were no more than shacks. The bus passed rice paddies with water buffalo working the fields.

There were poor little kids in the streets. As the bus passed by these kids, they would jump up and put their arms in the window while trying to sell some sort of baked goods. They would hang on to the bus as it moved through the town. The bus driver would not even slow down. I was afraid one of those kids was going to be injured from falling off the bus or under a wheel. When the driver drove into an intersection, he did not even slow down. He would just honk the horn and proceed forward at full speed. Full speed was probably about 45 to 50 MPH. People in the street would scatter for dear life. I had never seen anything quite like it before or since.

The Navy had arranged for a couple cases of beer to be on the bus. After we were about 50 miles up the road, sailors were asking the driver to stop for a pit stop. The driver would stop right in the middle of the road and say get out and go right here. The guys all lined up alongside the bus to pee. I thought this was crude, but maybe this is how they do things here. As the bus tour continued, I saw a woman squatting to pee along the road. Then I was convinced that when you have to go in the Philippines; you go right where you stand.

When we finally arrived at our destination, it was late afternoon. The place was nice. It was like a rural motel in any place back home. There was one difference; it was tropical. When we put our things in our room, I saw little salamander lizards all over the place. They had a supper planned for us. Then we walked up the road to a local bar and had a few beers. I thought if I drink enough beer, it would not matter how many lizards are in the room. Many of the guys used the swimming pool. We turned in early and got up early for the falls tour.

They next day we went to the river, and two by two got into a banca boat (dugout canoe) I called it a banana boat. Two banceros were strategically placed at each end of the boat, and we were paddled up the river

to the falls. This is the only access to the falls. On the way up the river, I photographed a local woman doing her laundry in the river. When we arrived at the falls, there was a little stand with a man selling picture post cards of the falls. This has been a tourist attraction since 1894. The falls is a very scenic place. Been there, done that, now I can check it off my list.

Later a Pilipino did a cockfight exposition and a barefoot man climbed a coconut tree. People were placing bets on the cocks, so I joined in on the fun and actually won. I was so into taking pictures that I did not even realize I had won. The honest Pilipino came to me with my winnings. I photographed a spider web that was about six feet across. The spider was at least six inches across.

On the return trip the bus went through Manila again. The tour guide was telling us there are extremely rich and extremely poor with no middle class in the country. From looking at the homes, it was obvious this was true.

As we passed by a cemetery, the tour guide told us that people are buried above the ground because of high water levels. It is the same as in New Orleans. One difference though: in the Philippines, there is an annual tax on the gravesite. If no one comes forward and pays the tax the body is taken out of the grave and discarded. The tomb is then re-used for someone else. This would most likely happen when there are no living relatives left. This was the custom in 1967; it may be different today.

Golf

The next day one of my friends from the supply department, Seaman Jack White, wanted me to play golf with him and Ron Smith. I was not much of a golfer, but dad had taken me to the Coal Creek Country Club a couple of times, so I at least knew a little about the game. The Binictican golf course on Subic Naval Base was designed by the Army Corps of Engineers in the 1950's. It was not real imaginative, but fairly easy and fun to play. Why not, the American taxpayers were paying for it. I had never played golf where monkeys were swinging from the trees in the rough. I heard horror stories where people on the golf course were attacked by monkeys. However, they did not look threatening to me. Jack and I told our caddie to catch one of them so we could take it back to the ship. He said no. We offered him $200 to catch one and he still said no.

I felt somewhat privileged being on the golf course, as the ship's captain was out there with us. Most of the people playing golf were officers. They were not in uniform, though. They were cordial to us.

A caddy was included as a part of the deal. He had made many rounds on the course and he had a good idea, which clubs to use at different spots. That was helpful.

At about half way through he said, "Today is your country's holiday."

I looked at him and said, "Really and what would that be?"

He said, "Thanksgiving."

I had no idea it was Thanksgiving. It was so easy to be wrapped up in the Navy's schedules that I would lose track of time and holidays. I was used to it being cold and wintery at Thanksgiving. Being in the tropics just was not the same. When we returned to the ship, the galley had all the Thanksgiving foods I was used to having at home.

As usual, I had taken my camera with me. When we reached the top of a hill there was a view of the bay with the USS Kearsarge sitting at the dock. From the tee box I took a once in a lifetime photo of the Kearsarge with a fairway in the foreground.

Chapter 10 - Port Of Sasebo

The Fire

December 23, 1967, the Kearsarge was entering Sasebo, Japan, for the Christmas holidays. I was standing in the hangar deck watching as the tugboat pushed her in. All of a sudden, the tugs reversed their propellers and started pulling the Kearsarge away from the pier. I had watched the ship come into port many times before, and I never saw them do that. I was wondering why, when I heard the General Quarters alarm. The ship was within about 50 feet of the pier when the GQ bell rang. That meant I had to leave the hangar deck and go to my battle station. The Captain announced there was a fire and when everything was secure, we would come back in. We went back out because they did not want to be sitting at the pier if the fires escalated and the ship blew up and sank. The dock would become unusable for a long time if that happened. There were family members standing on the pier waiting for us. I wondered what thoughts were going through their minds as we pulled away.

A fire in the O2N2 plant caused serious damage killing three and injuring two.

The fire was caused by a leak in a liquid oxygen line. Pure oxygen is highly flammable, so the smoking lamp was out due to the high concentration of oxygen in that compartment. According to Sonny Richardson, one of the men lit up a cigarette anyway, and instantly the compartment was engulfed in flames.

A twenty-two year old from Lebanon, Indiana, fireman DC3 Sonny Richardson, and Gary Vencill, as second hose man, were the first on the scene to put out the fire. The fire had gotten so hot it buckled the bulkhead. On the other side of that bulkhead were nuclear warheads. This could have been a very serious disaster. "SFM3 Harry Hope was the one that accidentally ignited the fire, and was one of the two that I had to carry out," said Sonny. Sonny carried two bodies out of the compartment. They were burnt so badly that dental records were used to identify the bodies. AN Robert Yoder and AK1 Kenneth Edwards also died in the fire. Sonny provided me a copy of the report he sent to the captain. Sonny was awarded the RV Gallantry Cross Medal which is awarded to persons who fulfill, with distinction, obligations above a level that is normally expected.[1]

When an airliner crashes, and a hundred people die and the news says ten of them were Americans, you have a feeling of sympathy for the ten Americans but have little concern about the other 90. That is because the Americans are part of us, and we do not know the other people. There were 4000 people on the Kearsarge and I knew only about 100 or so of them. When someone on the ship would die, I had the same feeling as when Bob Martin died. He was one of us. I always felt bad when someone connected to the Kearsarge would die even though I did not personally know him.

Rocking The Ship

While we were docked in Sasebo, Japan, Marty Sterrett and Ziebarth asked me if I wanted to climb the mast. I told them I would, as it sounded like an adventure to me. I had never been up there before. They didn't say why we were going up the mast and I did not ask. The yard arm is the cross section at the top of the mast. Before anyone can climb the mast, all radar must be shut off. If the radar were on, you would be cooked from the inside out, just like a micorwave oven. Since we were in port, all radar was already off. We didn't need to get permission to shut it down. We did need to make sure no one would turn it on while we were up there. The Navy had a safety system in which you would apply a padlock on the on-off switch so no one could accidentally turn it on. It was procedure to have our division officer and the OOD (officer of the deck) to both sign an authorization chit (a signed form giving permission) before anyone was allowed to go on the mast. I don't think we did that. The radar people and the electronic technicians were the only ones with keys to the hatch and the switch. We secured everything and started up the ladder. When I got about half way up the ladder, I looked around and discovered we were very high in the air. The hangar deck was about 30 feet from the water and the flight deck was another 30 feet above that. The super structure was about six levels, which would be another 60 feet. The mast was at least another 60 feet in height. So when you were at the top of the mast, you would be about 180 feet from the water. At that point I stopped climbing, but the other two encouraged me to continue. I wasn't sure what our mission was, and I thought they probably needed me so I continued all the way to the yard arm. When I arrived they were both standing on the yard arm. They started to sway back and forth and encouraged me to do the same. Their goal was to rock the ship from the top down.

They were pumping with their arms and legs the same way we did as a kid on a swing. Now I never would have believed it would be possible to rock that 31,000 ton ship with just three men that weighed only about 150 lbs. each. But, in fact, that is just what we did. The entire ship was rocking from side to side while sitting in port. My first thought was that we were going to be in deep doo doo for this one. As soon as they proved it was possible, they started to retreat down the ladder. We locked up the hatch, unlocked the switch and high tailed it out of there, and laughed about it for the rest of the day. The top of the mast was moving two to three feet from side to side. I wonder what the people who where inside the ship thought was happening as the ship rocked from side to side while in port. At lower levels the movement may have only been inches. About two thirds of the crew were on liberty at the time, so they missed it. The captain was probably not on board at the time either. It was an adventure that I will never be able to duplicate again in my entire life. That little event surely says a lot for the power of leverage.

Nagasaki Tour

The Navy arranged a tour to Nagisaki and I was eager to go. I wanted to go to the place where we dropped the atomic bomb to see for myself what we had done to Japan. I was now going to see what my mother was telling me about in 1952. The first stop was at the Saikai bridge. The tour guide pointed out the whirlpools swirling under the bridge. During my research for this book I zoomed in from Google Earth, and it is still swirling today. The trip to Nagasaki from Sasebo is about 40 miles.

On the way we stopped at a tortoise factory in which they were making products from tortoise shells. The products included jewelry boxes, combs, brushes and model sail ships. I believe it is illegal to sell or own these products today because we are trying to protect the sea turtles.

The next site we saw was the radio tower from which Tokyo Rose broadcast her propaganda messages to the US troops telling them we were losing and we should give up and surrender during WWII.

Our final stop was in Nagasaki. We toured a museum with artifacts from the city after the bombing. I saw a six pack of coke bottles that melted together into a glob of green glass. We went to ground zero where the atomic bomb exploded. Nagasaki was a target because it is where the one man subs were being manufactured and the Kamakasi Torpedo Factory was there.

At 11:02 AM on August 9, 1945 the United States dropped an atomic bomb known as Fat Boy on Nagasaki. It exploded 500 meters above the city. A heat wave reached several thousand degrees. Deadly radiation blast winds burned and crushed everything in site and reduced the entire area to a barren field of rubble. One third of the city was completely destroyed. Seventy-four thousand

people were killed outright and vaporized in seconds. Seventy-five thousand were badly injured. Six million seven hundred thousand square meters were leveled. Eleven thousand five hundred homes were vaporized. An additional 1,300 homes were destroyed. Fifty-five hundred homes were badly damaged. Large numbers of people died from the affects of radio active poisoning for years afterword. Japan had the highest rate of leukemia anywhere in the world.[2] This was the worst act being perpetrated on any human being on the planet, ever. My mother was so right when she told me the atomic bomb should never be used again.

We traveled to a mountain top where the view encompassed the entire city. With this observation I am convinced we should refrain from the use of nuclear warfare. However, when our opponent is as determined to win as Japan was, I don't think there is any alternative to using it.

It was said after the bombing that vegetation would not grow for 75 years. It was only 22 years later when I was there and everything was growing. They were wrong about that one. The damages were enormous, but that is what it took to get them to stop the war. Japan surrendered the next day. They have rebuilt their country since and have been a peaceful one, too.

When we returned to Sasebo, I went to a bar to have a beer, and the girl who waited on me sat with me for awhile. She spoke very good English. She was born in 1941, the year the war started. I asked her what she thought about the U.S. dropping the atomic bomb on Japan. She looked at me as though she had no idea what I was talking about. I asked her if they taught her about it when she was in school and she said no.

Sasebo Streets

Raw sewage ran in a gutter between the sidewalk and the street. Yes, it stunk. If you slipped off the sidewalk, you would be standing in the bingi ditch. No wonder I had to get all those shots before I could get off the ship.

Run In with the Shore Patrol

One night I was in town with Ron Smith. As we left a bar where we had a few beers, the shore patrol came by. We were both feeling good. We started making fun of them. We were saying, "Ha, Ha, you have to work while we get to play." They turned around, put handcuffs on us, took us in the paddy wagon back to the base, and held us for about an hour in the detention jail. They really had nothing on us, but they terminated our liberty for the day by taking us back to the ship and giving our liberty pass to the Master at Arms. The next day we were given our passes back. I guess we messed with the wrong guys. We were not drunk; we were just having fun. The only good thing that came out of that experience was that it gave me this story to tell.

Chapter 11 - Pueblo, Korea, & Home

USS Pueblo (AGER-2) is an American ELINT and SIGINT (Navy intelligence) which was boarded and captured by North Korean forces on January 23, 1968 in what is known as the Pueblo Crisis. The capture, less than a week after President Lyndon Johnson's State of the Union Address and only a week before the start of the Tet Offensive and subsequent 11-month prisoner drama, were major incidents in the Cold War.[1]

Cdr. Lloyd Mark "Pete" Bucher USN and 82 of his crew were taken as prisoners and held for 11 months.

We were in Subic Bay after being in the Gulf of Tonkin for a month. We had just arrived, but about two days later, we were on our way to Sasebo, Japan. When we arrived in Sasebo, we took on ship's stores, fuel, and ammo for several days. The Kearsarge had loaded so much stuff that it sank ten feet. I thought to myself, "My gosh. We just came from a war zone and we did not have this much stuff on board. What are we getting into?"

The ship went to the Sea of Japan and sailed in figure eights for a month. Russian ships were there with us doing the same thing. I took pictures of them and they took pictures of us. Nothing of any significance happened. A month later, we went back into the port of Sasebo and off loaded the ammo we had loaded a month earlier.

We were awarded the Armed Forces Expeditionary Service medal for our time in Korean waters.

The Long Journey

Just out of curiosity, I used Google Earth to see how far I traveled in 1967 and 1968. From June 25, 1967 to April 6, 1968, I traveled over 101,000 miles. One of the reasons the miles traveled are so high is the ship does not stop while at sea. I estimated about 400 miles a day for every day at sea.

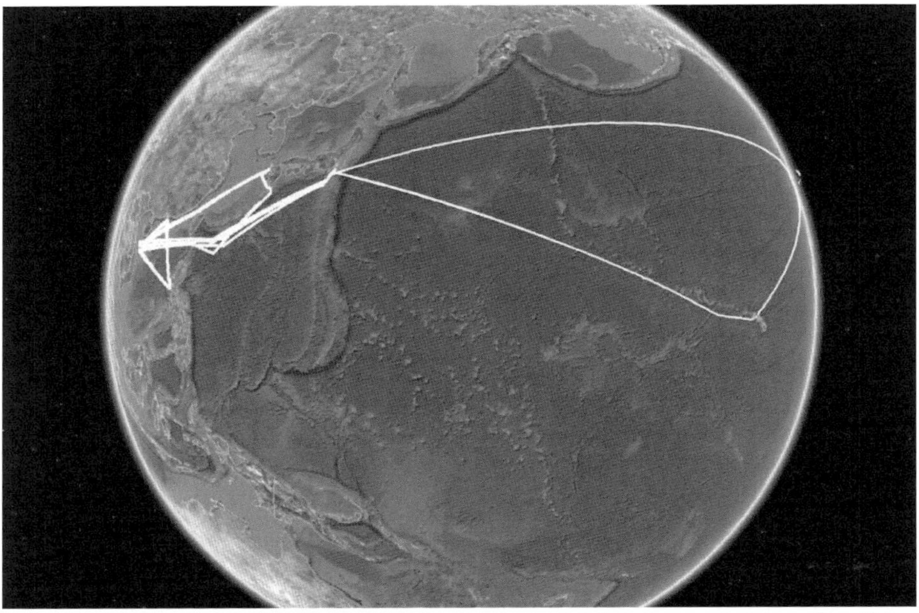

When traveling west, about every 24 hours we crossed a time zone and set the clocks back an hour. Roughly 36 hours after we departed from Pearl Harbor, we crossed the International Date Line and lost a day. It went from Saturday to Monday. On the return trip, we had two Saturdays in a row.

WESPAC Cruise from 1967 to 1968

10 Jul 67: We sailed for nine days off the coast of California for a shakedown cruise.

17 Aug 67: We started the WESPAC cruise; we left from Long Beach, California and sailed to San Diego, California to load the aircrew. Shortly after we left port, the airplanes followed and landed on deck.

27 Aug 67: Ten days later we stopped at Pearl Harbor, Hawaii.

29 Aug 67: We went to sea for three days and played war games in preparation for Vietnam.

3 Sep 67: We arrived back to Pearl for five days.

6 Sep 67: We sailed to Yokosuka, Japan.

26 Sep 67: We stopped for two days in Yokosuka.

28 Sep 67: We left port and sailed toward Midway Island chasing a Russian sub.

8 Oct 67: We were back in Yokosuka, Japan for ten days for repairs.

18 Oct 67: This was my first time in the Gulf of Tonkin, Vietnam. I was officially at war.

14 Nov 67: We were in Subic Bay, Philippines for supplies.

24 Nov 67: We were in the Gulf of Tonkin, Vietnam.

23 Dec 67: We spent seven days in Sasebo, Japan for supplies and repairs during Christmas week.

3 Jan 68: We were back in the Gulf of Tonkin, Vietnam.

19 Jan 68: We spent six days of R and R in Hong Kong.

26 Jan 68: We went to Subic Bay, Philippines for four days.

31 Jan 68: We were back in the Gulf of Tonkin, Vietnam for 21 days.

21 Feb 68: We spent seven days in Sasebo, Japan for supplies and repairs.

25 Feb 68: We were in the Sea of Japan off the coast of Korea for 26 days.

22 Mar 68: After three days in Sasebo, Japan for unloading supplies and ammo, then we sailed for San Diego.

6 Apr 68: We spent four hours in San Diego, California to unload aircrews, and that afternoon we arrived back at Long Beach, California.

Vietnam Poem

Author Unknown

Vietnam, Vietnam, another month and that's where I'll be. Wishing Lord that I could stay in the USA, Good old USA.

See that airplane in the sky. It's taking men over there to die. Wishing Lord that I could stay in the USA, Good old USA.

Over hill over dale hear that trusty MIG wail, but there is no end when Hanoi says to go to Hell.

Vietnam, Vietnam, nothing in that God damn country for me. Wishing Lord that I could be way back in the USA, Good old USA.

Chapter 12 - Dry Dock Overhaul-68

The Roommate

The Kearsarge had just returned from a tour in the Far East on April 6, 1968 and went into dry dock almost immediately for an overhaul. I was transferred to the barracks at the Long Beach Navy Station to help the supply department inventory the ship's electronic parts supply. I was attached to the supply department because they needed someone with electronics experience to help identify items that have lost their tags. This job lasted all of the summer and part of the fall of 1968. It was like a civilian job in which I worked from 08:00 to 16:30 with weekends off.

The rooms in the barracks had two single beds and a closet. My roommate was Robby Hiker. He was from a wealthy family in Arizona. He told me his mother was politically connected. She told him Richard Nixon would be the next President. I asked him how he could be so sure and he said, "Just mark my words, it will happen." It was not long after that that Bobby Kennedy was assassinated. Later that year Richard Nixon was elected President.

Robby always wanted to borrow my civilian clothes. I did not mind if he would ask first, but he started wearing things without asking. When I would look for a shirt, it would not be there. It got out of hand and I had to cut him off. He also was always asking to borrow money. I was reluctant to loan money to sailors. The ones who always-borrowed money just spent too much and were always broke. My thought was that if you do not have money now, how are you going to pay me later? How is it that a man from a wealthy family is always broke?

One day I gave in and loaned him $20. Later that day he wanted to go in town with me. We stopped at a bar and had a couple of drinks. He said, "I'll buy." I said, "Okay." Later when he was paid and I hit him up for my $20, he told me he did not owe it because he had paid for my drinks. I never loaned him any money again. Some people take advantage of others, not necessarily intentionally, but, nonetheless, they are not to be trusted. Robby was one of those people.

Promotion to ETN3

Shortly after returning from Vietnam on April 16, 1968, I was promoted to Third Class Petty Officer (E-4) which raised my pay to $234.00 per month.

The Old Salt

When the ship returned from our WESPAC Cruise, lots of changes were taking place. There was a man who was at least 50 pounds overweight. I would see him from time to time sweeping the hangar deck. I guess the reason I saw him in the hangar deck all the time is because his sole job on the Kearsarge was to keep the hangar deck clean. I would guess his age to be about late thirties or early forties. He had been in the Navy since he was 18. He was not very aggressive because his rank was Third Class Petty Officer. Considering all the years, he should have been a higher rank. I believe he was in the boatswain department. A year earlier, he was told by the ship's doctor that he needed to lose some weight.

The man would collect old toothpaste tubes from the trashcans in the head and sell them for the lead. (In those days tooth paste tubes were made from lead.)

One afternoon when I was leaving the ship, I saw him sitting on the dock with his sea bag. A few days later, he was still sitting on the dock in the same spot, crying. Because of his weight and his low rank after twenty years of service, the Navy gave him a discharge. The only thing this man knew was the Navy. He had been on the Kearsarge for the past eight years. He never got off the ship at any of the ports. I was told he had no family. He really had nowhere to go. He did not want this discharge. The Navy took him to a home for old salts with nowhere to go. I felt sorry for the man, and swore that I would never end up like that.

July 4, 1968

At the end of June 1968, my dad called me at the barracks, and wanted to know if I could come home for Independence Day weekend. I told him I did not know if I could get the time off and that I could not afford the flight. He said that if I could get the time off, he would pay for the flight. He said that Larry was going to be home and it would be nice if I could be there for a family reunion too. He said Larry would ship out to Vietnam afterward and it would be nice to have the entire family together. The fourth fell on Thursday and I needed Wednesday and Friday off, as I already had Saturday and Sunday off. This would give me five days for the trip. I put in a request for a couple extra days during the holiday weekend and my superior, Second-Class Petty Officer Billings okayed it. On Wednesday, I flew military standby on a flight from LA to Chicago for $50 round trip.

While at home, we had a family picnic at Johnson Sauk Park in Kewanee, Illinois. My brother-in-law brought his rowboat and we all took turns going out on the lake. We drank beer and Dad cooked burgers and hot dogs on the grill. It was a great Fourth of July. Later in the evening, all of the older kids went to a bar and had a few drinks. When we returned, Larry and I sat in the living room and talked until

04:00. One of the things we talked about was, "being confronted one-on-one with the enemy." I asked Larry if he could shoot and kill someone. He said he did not know. This thought has haunted me ever since. I did not know when I left the house that Sunday morning that it would be the last time I would see Larry alive.

When I returned from my long weekend at home, Petty Officer Billings asked if I had a nice weekend.

I said, "Yes."

He said, "Where did you go?"

I said, "I went home."

He said, "Where is that?"

I said, "Sheffield, Illinois."

He said, "You were not supposed to go any farther than 50 miles from the Navy base."

I said, "Well, I went 2000 miles to see my brother who is shipping out to Vietnam."

He said, "Okay, but I would not have approved it if I had known you were going that far."

A Job Offer

On my return flight to Long Beach, I sat next to Richard Ireland. Richard asked me what I did for a living. I told him I was an ET in the Navy. Richard worked for Smith Corona, the typewriter company. He told me if I needed a job when I got out of the Navy, to give him a call. I always wondered how that would have turned out if I had pursued it. By the 1980's the new word processors on computers caused the typewriter to be placed in a museum with the buggy whip. If I had to write this book with a typewriter, it never would have happened.

Met With Lynn and Steve

In the summer of 1968, Lynn VanAcker was back in Long Beach. He and I spent quite a few days together. We were walking on the sidewalk in Long Beach when a moped came roaring up the street. It was Steve Spratt. What were the odds of the three of us running into each other when we all had been assigned different ships with different schedules? We spent that afternoon together at the beach.

We all decided to get a tattoo. Steve had a date, so he had to leave. He said he would meet Lynn and I at the tattoo parlor at 19:00 in Long Beach. Lynn and I showed up, but Steve did not. Lynn and I decided not to get a tattoo until Steve was there. We were all going to get the same thing. In the end, Steve got one and Lynn and I never did. I have always been glad Steve did not show up. I would have regretted getting the tattoo.

One night Lynn and I were having a cup of coffee at the Greyhound bus station. A very attractive girl walked by the window and looked at the two of us as she went by. We both commented about how nice she looked. About a minute later, she came back and walked up to us and asked if she could join us. We said yes. She said her name was Sherry. After talking for a while, she invited us to her apartment. We have no idea where she took us, but it was a long ride from the bus station. It was late in the night and she was showing Lynn her high school albums. At about 02:00 Lynn laid down to sleep. She continued to talk to me until the sun came up, about 04:30. She told me she had to go somewhere and would be back in a couple of hours to get us. I said, "Fine, I am going to sleep on the couch." About 07:30, Lynn woke me up and asked where she had gone. I told him that she should be back anytime. We sat in her apartment until almost noon. We would have left, but we did not know where we were. Lynn said the closet door was open and there was a Navy uniform hanging in the closet. She told us she was not married, but we could see she was seeing someone who was a higher rank than we were. She finally came back about 13:00; she wanted to take us to another place to party. We told her to take us back to the bus depot, so she did. We both thought the whole thing was bizarre.

A few weeks later, I saw Sherry in a bar in Long Beach and she said hello to me, but I acted as if I did not know her. She was hanging with a group of five sailors.

Summer of 1968

In the summer of 1968, my sister Becky came to California to visit. We went to Disneyland and Marine Land. She had a male friend who took Robby Hiker and me out to lunch and to an afternoon of playing pool down by Newport Beach.

Bob Mahara took me to see the play "Hair" in Los Angeles. He was good to me. He took me to a party with all doctors and nurses. The conversations were strange. They talked about playing rock-and-roll music in the operating room during surgery. They also talked about doing surgery on a fat person while balls of fat were dripping on the floor, and they were slipping and sliding in it. He once took me to a party in an elite part of Hollywood. He was sure that Raymond Burr from the TV show, "Perry Mason," lived in the house across the street.

In the first week of October, Chris Sandburg, one of the girls from Princeton, Illinois that I used to hang with, came to Gardena, California and stayed with her aunt Hazel. She invited me to stay the weekend at her aunt's. She and I went to Disneyland the next day. It was interesting to be at Disneyland at that time of the year. It opened at 08:00 and closed at 17:00. There were not any lines to any of the attractions. We went into everything in one day. It was a fun day. I never saw Chris again until Larry's funeral in February 1969, and I have not seen her since.

Downtown Long Beach

To occupy my time when on liberty, I would go to the United Artists movie theater in Long Beach. It was a different kind of place from any theater back home. It was open 24 hours a day. They would show a different movie every couple of hours. You could buy a dollar ticket and stay all day. Sometimes I would go there and watch three movies in a row. The movies were not always current, but they were not old ones either. They ran many documentaries, which I enjoyed watching. I had a military pass that allowed me ticket discounts.

There was an amusement park called the Pike down by the beach. It had the typical stuff: a roller coaster and Ferris wheel, a midway and bars, tattoo parlors, and nightclubs. Many sailors went there for something to do. I liked to frequent one particular bar with live country music. All of that has been torn down since those days, but the street is still called the Pike.

We used to call Long Beach, the City of the Whispering Bushes. There were many queers preying on sailors. When walking through the city park, the bushes would talk. They would say, "Psst, hey buddy, come here." I just ignored them and they would leave me alone. I found it to be repulsive.

One evening I was on my way back to the Navy base, which is about three miles from the downtown Long Beach, and I decided to thumb a ride. A big white Cadillac stopped and a woman inside asked where I was going. I told her back to the Navy base.

She said, "I am going that direction, get in. I will give you a ride."

I jumped into the car, and she started to drive and I heard the doors lock. She asked me if I wanted to go to a party. Her voice sounded as if she was a man. She had feet that were bigger than mine were. She had hands like a man.

I said, "No thanks; you can let me out right here."

She said, "No, I promised you a ride to the base and I will take you there."

It only took about five minutes and we were at the base. She stopped at the gate and asked me one more time if I wanted to reconsider going to a party. I told her no and got out of the car. I never hitched a ride again. I had always taken the bus to and from the base, and I went back to taking the bus. I am sure she was a man who was a cross dresser. She turned around and headed back to where we just came from. Was that person a she, a he, an it, or just a shit?

I did a lot of walking in those days. I would walk up Magnolia Avenue, from Ocean Boulevard to Pacific Coast Highway and then to the traffic circle to a nightclub and then walk back. The round trip is ten miles.

One Saturday afternoon I was walking down Long Beach Boulevard with Robby Hiker. There were about 25 people kitty corner from where we were standing. About that time, three police cars and a paddy wagon pulled up.

Robby said, “Let’s go over there, and see what is going on.”

I said, “No, my dad told me to stay away from things like that.” I said, “If we go over there, we could be dragged into it.”

He did not agree with me, but he went along with me, and we stayed put. Sure enough, the police loaded every person who was standing on the corner into the paddy wagon and then took them to the police station. Score one for my dad.

Meritorious Unit Commendation

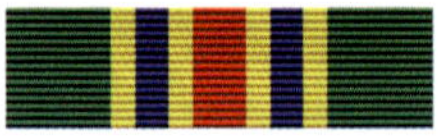

It was because of the 1967 – 1968 cruise that the USS Kearsarge was awarded the Meritorious Unit Commendation. On the afternoon of July 29, 1968, Captain Creighton W. Cook, Commanding Officer of the USS Kearsarge, announced that the "Mighty Kay" was to receive the third highest award given a United States Naval Unit.

Admiral T. H. Moorer, Chief of Naval Operations, speaking for the Secretary of the Navy, set forth the following citation:

The Secretary of the Navy takes pleasure in presenting the Meritorious Unit Commendation to USS Kearsarge for service as set forth in the following citation

For meritorious service while assigned to the Seventh Fleet and serving off the coast of North Vietnam and in the Sea of Japan during the periods 11 July to 11 December 1966 and 12 October 1967 to 28 March 1968. For providing outstanding command and control facilities for prosecution of all assigned tasks. USS Kearsarge deployed into the Sea of Japan in August 1966 where she operated with Japanese and Korean naval units and thwarted aggres-

sive harassment by foreign naval units. Kearsarge also conducted operations with the United Kingdom, Australian, and Chinese ships and provided support to the President of the United States during visits to Manila and Kuala Lumpur. During the 1967 deployment, Kearsarge, her air group, and destroyers in company demonstrated superior capability in maintaining contact with high speed, conventional submarines in the open ocean on two occasions while developing new and unique tracking techniques. While in the Gulf of Tonkin, Kearsarge displayed superior readiness by continuing around-the-clock aircraft operations without mishap during the northeast monsoon season. She also provided exemplary Surface-Subsurface surveillance; gunfire spotting for cruisers and destroyers; search and rescue support; daily mail, personnel, and priority cargo transfers; and responsive medical evacuation. The officers and men of USS Kearsarge made a most significant contribution to United States efforts in Southeast Asia. Their courage, skill, adaptiveness, and devotion to duty were in keeping with the highest traditions of the United States Naval Service.

Allpersonnel attached to and serving on board USS Kearsarge (CVS-33) during the above designated periods, or any part thereof, are hereby authorized to wear the Meritorious Unit Commendation Ribbon."

(Signed) For the Secretary,
T. H. Moorer
Admiral, United States Navy
Chief of Naval Operations

The unique aspect of this citation lies in the fact that the "Mighty Kay" is the first and only anti-submarine warfare support aircraft carrier operating in the Pacific or Atlantic fleets to receive the award. This achievement is another prime example of the proud heritage embodied in the motto of the USS Kearsarge: IN OMNIBUS PINNACULUM or, if your Latin is a trifle rusty, IN ALL THINGS THE BEST. The Ribbon is awarded to a unit for valorous or meritorious achievement in a combat or non-combat situation.

Laser Factory

In the fall of 1968, the Navy took a busload of ET's to a laser factory in Los Angeles. The laser was an experimental device at the time. The term "laser" originated as acronym for *Light Amplification by Stimulated Emission of Radiation*. It is an amplification of photons concentrating of light.[1] The lasers we were shown were six feet in length. There were many different concepts being developed at that time. They even had a beer laser. In a demonstration, they placed a quarter in a vice and shot the laser at it. The laser burned a hole in the quarter about the size of a human hair. It happened in just a second or two. Then they shot a hole through Lincoln's eye in a penny. It was a very impressive demonstration. It was just like the laser we saw in the 1964 movie, "Goldfinger."

The man who was showing us around told us that in 1969 when the astronauts would go to the moon, they would place a mirror on the moon. Then he said they were going to take this laser to McDonald Observatory and aim it through the telescope to measure the moon's wobble. Buzz Aldrin and Neil Armstrong put a 100-mirror panel there on July 21, 1969. With this experiment, they have been able to determine the moon is spiraling away from Earth at a rate of 3.8 cm per year. This is the only Apollo 11 experiment that was still in operation 35 years later.[2]

Lasers have come a long way since then. They are now in our DVD players and we use them as a pointing device. They are used in all kinds of medical devices today. It was fun being among the first people to be introduced to them in 1968.

San Diego School

October 19, 1968, I was transferred to temporary duty in San Diego for Electronic Test Equipment Calibration School. This was a six-week school. I really liked the class and learned a lot about calibration. This was one of those times when we were given the final test first and again at the end. At the beginning, I scored 40% and at the end, I scored 95%.

One day after a weekend of liberty, one of my classmates was telling me about this girl in Long Beach that went out with a friend of his and four other guys. He said she took them back to her place and had sex with every one of them. He started to describe her to me. I thought that sounded like the girl Lynn and I met. I asked if her name was Sherry. He said, "Yes do you know her?" I told him that Lynn and I had met her, but we did not have sex with her. He said, that it was a good thing we had not. His friend later ended up with gonorrhea and so did the other four. Sherry was a little whore spreading disease among the troops.

Before I left San Diego, Roy Moruss came to visit and showed me around Tijuana, Mexico. He had been there before and had special places he wanted me to see. My take on it was, it was a lot like Olongapo City, Philippines. They both qualify for the title "Sin City."

I finished the school at the end of November then returned to the Kearsarge. The yard period had continued from April to October 1968. The ship underwent sea trials on and off from October 1968 to February 1969.

Calibration School Graduation

Battle Efficiency Award

On November 13, 1968, I received a letter from Captain Cook. At first, I did not think much about it, but if you look at the Calibration School graduation photo, I am the only one wearing one. Some of those men had been in the Navy for 16 years.

Commander, Naval Air Force, Pacific takes pleasure in awarding the Battle Efficiency Pennant to the USS

Kearsarge (CVS-33) for standing first in Battle Efficiency Competition during the Competitive cycle January 1, 1967 to June 30, 1968. In addition, the USS Kearsarge (CVS-33) is awarded the Operations "E" and Weapons Black "W" for standing first in the Operations and Weapons excellence Competition during the Competitive cycle January 1, 1967 to June 30, 1968.

As a member of the Ship's company during this period, you are commended for your exemplary performance and consistently outstanding team effort that produced these awards. The perseverance, skill, and dedication displayed by you and your shipmates have directly contributed to this recognition of superiority in safety, efficiency, and combat readiness.

C. W. Cook
Captain, U.S. Navy
Commanding Officer

Sheila, My New Girlfriend

It was late afternoon in early December 1968, when I met Sheila in Long Beach. I was sitting on a bench in the city park when I noticed a very attractive lady walking with her three kids. She was talking to her youngest girl in sign language. I commented about how amazed I was that the little girl knew sign language. Sheila was a petite, blue eyed, blonde lady with whom I had an instant attraction. She was speaking to the two oldest kids in a foreign language. I asked her what it was and she said, "Dutch." She was speaking perfect English to me. As we talked, I learned that she spoke Dutch, English, Spanish, and sign language. She told me she was a divorcee. She said she lived in an apartment on Magnolia Avenue just a few blocks from where we were in the downtown area. As we talked, we discovered we shared the same birthday. I asked her out for a date and she accepted. That date led to us to becoming a couple and we spent every day together when I was off duty.

Christmas Leave 1968

December 13, 1968, I flew home for a 14 day Christmas leave. My brother Wayne was still in the Air force and could not get home, and my brother Larry, who was in the Army, was now in Vietnam. My friend Lynn VanAcker was home and he came to the family house to visit. The trip home was a good one. When I returned to Long Beach, we would be preparing for another trip to Vietnam.

Long Beach Party

On New Year's Eve, the ship hosted a party for the ship's crew at the Long Beach arena. I went to see what was happening there. They had lots of food and free drinks. They had a five-foot ice carving of the ship on display. That was cool! Literally!

We were told to have a good time but not to get drunk and disorderly. I stayed for only about 30 minutes because I did not see anyone from my division. As I was leaving, I saw a drunken boiler-man starting to pick a fight with another sailor. Against my better judgment, I went over to break it up and he punched me in the jaw for interfering. At that very moment, the shore patrol came around the corner and saw his punch. They wanted to know what was going on. I told them I tried to break up a fight. They handcuffed both of us, threw us in the paddy wagon, and took us to the holding cell on the base. The guy he was fighting with walked away. Because my jaw was bleeding, I was considered guilty of fighting. They pulled our liberty passes and took us back to the ship.

I never got involved in anyone else's issues again. This is the reason people turn the other way when someone is in distress. If you get involved, it may become your fault. The Navy taught me well. My dad was right- walk away.

New Commanding Officer

On January 17, 1969, Captain Leonard M. Nearman became the twenty-third and final commanding officer of the USS Kearsarge. He showed considerable concern for the morale of the crew. They in turn dubbed him "Top Cat."

Chapter 13 - Larry's Death

The USS Kearsarge had been at sea for just one day off the coast of California. We were working near Santa Catalina Island. I had taken a picture of the island on February 3. We were going be out for only a few more days doing exercises in preparation for another trip to Vietnam.

On February 4, 1969, at about 19:00 hours, the Chaplin called me to his office. I had no idea why he wanted to see me. My first thought was the Navy is going to recruit me to attend Sunday church services. I had visited with the Chaplin before about attending services, but I had not followed through with it. The Chaplin told me to sit down. Then he read a telegram from the American Red Cross in Washington DC to me. It said, "On February 3, 1969, Larry R Maddy had been killed in action in Vietnam." That was the last thing I expected to hear. I was in shock; I did not know what to think or say. I wanted a re-wind and to come back in again with different news. I just sat there staring at the floor. The Chaplin asked if I was okay. I said, "I don't believe it, are you sure you have the right person?" The Chaplin read the telegram to me again. This time I started to cry. The Chaplin just comforted me as best he could. After I calmed down, he told me I could go home the next day. I said "How, we are at sea?" The Chaplin told me he would arrange for me to fly back to Long Beach in the morning. He told me he would be available for me anytime if I needed anything else. I went back to my quarters and started packing some clothes for the trip home.

The next morning at 05:10, I boarded a transport plane and catapulted from the Kearsarge. They flew me into the Navy base at Los Alimotis with instruction to call home when I landed. From there I taxied to Sheila's home, and she drove me to LAX. I flew home to be with my family.

I did not think about it at the time, but I had just been launched off an aircraft carrier. I wonder how many people in the world get that opportunity. Most people in the Navy never get to do that, not even those who are stationed on a carrier get to do that. I was seated facing the back of the plane and strapped in like a race car driver. I was in a 12.5 ton airplane that was about to take off from a 150 foot runway. At takeoff, there was a huge thrust as the catapult hurled the plane in the air. It slammed me back in the seat and I felt as if it was wrapping my stomach around my spinal cord. Three seconds later we were airborne. Once the plane was off the ship, it dropped a few feet. It was almost like being on a roller coaster. Even though I had just lost my brother, I had also been blessed with an experience that I will never have again. I was not thinking about it while it was happening, but I sure did think about it later. Less than .0006 percent of America's population has ever flown off an aircraft carrier. Only about .6 percent of those in the Navy ever get to fly from a carrier.

The plane flew over the ocean. As I looked out the window, I could not tell how high we were because there was nothing below to gauge the height. I had never flown over the ocean before. After about 30 minutes, I saw the California coast; it turned out we were not very high at all. Those 30 minutes seemed like hours at the time.

I started thinking about what my mother had said back in 1952 when she and I were sitting on the couch. She hoped that her boys would never have to go to war. Now she had a son who was killed in action. I thought about the night that Larry and I talked about whether he could kill another person. I kept thinking this was all a dream and that nothing was real. I felt that if I could just get home and see my family, they would tell me that none of this was real. The flight to Chicago was a long one. I just could not seem to get home fast enough. Every time I would think about what was going on, I cried. I tried to occupy my thoughts with other things, which was not easy. Maybe I should have shot his eye with the BB gun when we were fighting. If I had, he would not have been eligible for the Army and would still be alive today. If I had done that, then I would have always been the bad person for doing it. Now I would never have the opportunity to have him pay his debt for the car he wrecked. He would never be married nor have any children. He would not get to grow old and reminisce about the old days. I kept wondering about my mom and my dad. They always seemed to be able to fix things that went wrong. What were they going to do to fix this?

When I arrived home, my dad was strong. He said, "Hold your head up high and be proud and be strong like the Kennedys were." He said, "If you hold your head high, you won't' cry." He was right; that worked. However, it did not make the pain go away. I thought, "Wow, my dad is comparing us to the Kennedys." The truth

is that it does not matter whether it is a fallen President or a fallen soldier; the family still has the same grief and has to deal with it. Just a little over five years earlier John Kennedy had died.

LT JG O'Donnell told me when I left the ship that I could be gone for two weeks, but if I needed more time I could just call, and they would extend it. After the funeral was over and everyone had gone home, my dad asked me when I had to go back. I told him I had two weeks, but the Navy would extend it if I needed to. He asked me to extend it to three weeks. I do not think he wanted me to leave. By this time Wayne had been discharged from the Air Force, and I was now the only one left in the military. When I was leaving Chicago to return to the Kearsarge, my dad kissed me goodbye and told me he loved me. That was the first time he had ever done that. I was 21 years old.

My dad was strong while I was still home, but he and my mom had a hard time holding it together after all the press and family were gone. No one should lose a child this way.

When Larry was in high school, he wrote a paper on the great documents of freedom, the "Declaration of Independence," the "Constitution" and "The Bill of Rights." He went into the military and took an oath to defend those documents. If he were alive today, he would be appalled on how these things are being trampled on by our politicians who took the same oath to protect them.

When I returned to the Kearsarge and told second class Petty Officer Billings that my brother Larry had been killed in Vietnam, he was glad he approved my long weekend leave back in July. When I first met Billings, he came across as a tough guy, but he really had a soft spot in his heart.

The men in OE division gave me their condolences when I returned. One man in particular, Richards, was quite sympathetic and told me anytime I needed to talk he would be available. After losing Larry, I became very cynical about the military. One day when I was talking to Richards about how I felt, he told me to knock it off. He was the least sympathetic guy I had ever met. He said it was okay to die for the country and it was Larry's turn to die. I never spoke to

Richards again. That was not what I needed to hear just three months after Larry died.

My dad wanted me home. The family started a letter writing campaign to get me released from the Navy. There were letters from the church pastor and doctors stating my dad's mental stress from Larry's death. I had written letters to congressmen, and Senators. It was a long and tedious process, and we were getting nowhere. In the middle of my second cruise, the Kearsarge was ordered back to Long Beach. My time left in the Navy was short and I just gave up on getting out early.

Chapter 14 - Second Vietnam Cruise

My New Job

During the last cruise, I was the parts supply man for OE division. After having completed the calibration school, I was placed in TGR (test gear repair) for this cruise. My job included repairing equipment and doing an accounting for every piece of equipment on board the ship. Hundreds of volt/ohm meters, signal generators, and oscilloscopes were located in compartments all over the ship. During my inventory count, I discovered that 100 pieces were missing and unaccounted for. They were most likely stolen by sailors who had been discharged from years past.

Every time we entered a port, I would gather equipment that was not working or that needed to be calibrated, and take it to the base electronic shop. It was my responsibility to get it there and back while we were in port.

Back at Sea

On March 29, 1969, I was at Sheila's apartment when I awoke at 05:45. I had fifteen minutes to get to the ship. An officer picked me up and gave me a ride to the ship. I made it with five minutes to spare. There was very heavy fog overcast at 08:00; the ship did not leave as scheduled. At about 08:30 the Master-at-Arms made everyone who was not in dress blues leave the weather decks; that included me. I guess they wanted to take pictures. At 10:00, I went to the shop to see what work had been done on the signal generator I had been working on. At 11:00, I went out on the weather deck and looked to see where we were. We were just outside the break waters of Long Beach. The USS Kearsarge departed Long Beach on the start of her Eleventh Far East Cruise and steamed to San Diego to pick up the air group at North Island Naval Air Station.

Then I hit my rack until 19:30. When I woke, we were in San Diego. As soon as the air group was aboard, the ship would leave San Diego and steam to Pearl Harbor. Four other guys and I decided to go to the E.M. club (enlisted men) and have a beer before we left the states. We had to be back by 21:30 because the ship was leaving at 22:00. My first ambition was to call my sister Becky, seeing how I forgot to the night before and had promised I would call before we left the states. After the beer, I went back to the ship and hit my rack again. The ship left port at 22:30 and shortly after, the airplanes landed on the ship. There were about 40 aircraft.

It seemed as if all night long they were paging people on the ship's 1MC. All the noise kept me from getting a good night's sleep.

On this cruise, there were 30 men in my division and only eight of them were on the 1967 cruise. Seventy-three percent of the men were new to the Kearsarge. That is the reason I said it is amazing the ship functioned as well as it did.

The Navy increased my pay to $268.00 a month during this cruise. One afternoon we were talking about how little we were paid and one of the guys figured it out to be .37 cents an hour. Then he laughed and said that the good news is you are paid 24 hours a day. He said where else could you get a job where you are paid even when you are sleeping. We also were being paid an additional $10.00 for sea duty and $65.00 for combat pay, giving me $343.00 a month. The good news was that what we earned while in combat was tax-free.

Two days out of San Diego Captain Nearman announced that he hoped the cruise would be a pleasure for us and that he would do his best to make it that way.

Tora! Tora! Tora!

On April 4, 1969, I woke up at 07:00. The ship was already in sight of the beautiful state of Hawaii. With all of the excitement of getting into port, I did not bother to eat breakfast. At 07:30, we began to enter Pearl Harbor. The filming of the movie "Tora! Tora! Tora!" began at about this time. Twentieth Century Fox was making a movie about the bombing of Pearl Harbor. The ship was to be filmed entering Pearl Harbor at dawn. We arrived at Hawaii on the afternoon of the day before. We stayed out to sea all night so Hollywood could film a carrier entering port at dawn. I was not happy about it; I wanted to get into port. The USS Kearsarge played the part of the USS Enterprise as she entered Pearl Harbor after the Japanese attack. The ship's number "33" was painted over so it would not show up in the movie. There were two cameras, one on the shore and one in a helicopter. In the harbor, they had imitation sunken battle ships and made up after effects of the bombing. Hollywood had done such a good job with all the props that it really did feel as if we were arriving into Pearl Harbor right after the bombing. There was smoke and fires were burning all over. There were soldiers who were dressed in WWII uniforms manning machine guns behind sand bags along the

shoreline. There was a full size mockup of the battleship Nevada at Ford Island, and three Japanese zero aircraft were flying in the area as we entered the harbor. It was the last scene in the movie. They filmed us for 15 to 20 minutes as we entered port and used two to four seconds of it in the movie. By 08:30, the filming was complete and we were in port.

The XO (executive officer) said that liberty would start at 16:00. The CO (commanding officer) said no, there was too much to see and it couldn't be seen inside the steel compartments. He let us go at 13:00, three hours earlier. Before going on liberty, I had test equipment to deliver to shore station calibration. By 14:30, I was in downtown Waikiki. Roy Moruss and I checked into a Motel 6, then went for a walk about town and took pictures. At 17:00, we went swimming at Waikiki Beach. In the evening, we left to go eat and I ran into Bob Maehara. We ate at the Rigger and later went to a topless nightclub. The night ended watching TV in the motel at 01:00. Roy and I rented floor space to about eight other sailors for the night. We had so many men in the room, the floor was wall-to-wall men, and John Hand ended up sleeping in the bathtub.

Re-Occurring Nightmare

After Larry died, I had a re-occurring nightmare that I was in combat in Vietnam. As we came into a town, we were told to search the huts. I saw someone run into a hut, so I followed him there. As I entered the hut, the man who looked Oriental was speaking English. He said, "Don't shoot; I am one of you." I did not know what to believe. He was dressed like the Vietcong. I did not want to take a chance at being shot. He was pointing a gun at me and I was pointing a gun at him like a standoff. I was thinking about my conversation with Larry about shooting or being shot. I pulled the trigger and shot this man, as he said "No." Then I would wake up from the dream. The nightmare went the same way every time I had the dream.

In April 1969 when we stopped in Hawaii before going to Vietnam, I told Dr. Maehara about my dream. When I said I shot the man because he was Oriental looking, Bob looked at me and said, "Then you would have shot me?"

I had not thought about it, but I said, "If I didn't already know you, yes, I guess I would have." I felt sick about my answer to him. He was disturbed by it also. I told him I could not help that it was that way. It was just a dream.

After that while I was in Vietnam, Dr. Maehara was drafted into the Army, but by September he was out already. When I asked why he got out so quickly he said, "I will tell you later." It never came up again. I often wondered if my nightmare had anything to do with it. I knew it really bothered him. I never had that nightmare again after I told Bob about it.

I had lost a lot of weight already this cruise. I just did not have an appetite. I had been quite disturbed over Larry's death. I could not forget it after being around officers and hearing news of the war every day and being involved in the war. My division officer kept showing propaganda movies about war and the reasons it is necessary. The officers in CIC seemed as if it were a real thrill to be in war and kill other human beings. It just made me sick to the point that I could not eat or even sleep a lot of the time. I refused to watch any war movies on TV, I tried not to hear the news for the day, and I tried to ignore the officers and their efficiency war wishes. All of this had made me so depressed that the only thing I had was mail. I knew that mail from home always had a good word and moral support. I really felt bad about not writing home often, but there was nothing to tell, except gloom and doom. I did not want to tell them that. I saw no reason for this war to go on.

Underway Again

On April 7, 1969 I got up at 06:30. I needed to retrieve my laundry from the pier before the ship left Pearl Harbor. When Liberty call expired, it was funny watching the guys return to the ship: they were drunk, beat up, wet from jumping off the pier, and decorated with leis.

I spent the day setting up a file for the tech manuals of the test equipment. For the past couple of nights I had been cleaning out a shelf or two at a time. Anything that was not accounted for or had not been used for a year and would not be used for a time, we had thrown overboard. This really amounted to quite a lot of stuff.

I was very tired because I got very little sleep the night before. I quit working at 17:00. I did not eat because I had no appetite.

After our movie debut, we operated on operational readiness exercises in Hawaiian waters for about a week, and then departed for Yokosuka, Japan.

On the April 8, 1969 at about 02:00, the ship began to roll very badly. It woke me by almost rolling me from my rack. It was the worst rolling I had ever experienced so far. I just laid flat, grabbed the edges of my rack, and tried to get back to sleep. I woke with a bad headache and sick to my stomach. At noon when I went to chow and the ship's band was playing in the hangar bay, which was very nice. All day long, the sea was rough with water splashing into hangar bay three.

A few days later on the twelfth, I had a miserable night. Taps went at 22:00, and immediately there was an exceptionally loud clatter on the side of the ship. It lasted about an hour. This happened every night at the same time when we were at sea. I hated it, but I guessed there is nothing I could do about it. At 04:00, I had to get up for the 04:00 to 08:00 watch so I really needed to get some sleep.

With some of the activities going on you might have thought we were on a civilian cruise ship. The next day at 13:00, the ship's band had a concert in the hangar bay three. They were good and I really enjoyed it. The captain authorized fishing from the fantail until flight operations needed to resume. It was different. I did not fish, but I watched. No one caught anything. Fishing from the fantail with four ten foot screws turning the water - what are the odds that you would catch anything?

May 3 was an unusual day. In the afternoon at 13:30, the CO called away holiday routine. Along with it he authorized sun bathing on the flight deck, bingo on the forecastle, movies in the amphitheater, volley ball in the number one elevator, ping pong in the mess hall, basketball in the hangar bay number three and at 17:30 supper was served picnic style on the flight deck.

In a couple more days USS Kearsarge was "On the line in '69" operating on "Yankee Station" with liberty periods being spent in the same ports as the 1967 cruise: Yokosuka, Sasebo, Subic Bay, and Hong Kong.

Another Broken Heart

I was still dating Sheila when I went home for Larry's funeral in February 1969. I told my sister Vivian I thought I would marry her someday. When I returned to Long Beach, my mother wrote me a letter expressing her disapproval of me being involved with a woman who had three kids when I was so young and had a full life ahead of me. I was disappointed by how mom felt, but I understood.

Later in May 1969, we were at sea on our way to Vietnam and I overheard Gary Cramer, a man in my division, telling another man about a woman who was having sex with a group of men on the living room floor. Then he described her and said her name was "Sheila." My heart fell to my stomach. That was it for me; I was done right there. This apparently happened while I was home for Larry's funeral. She wanted to get married so she could get spouse pay from the Navy and if I were to die while at war, she would get benefits without the attachment to me. She had said something about that at one time, but I just ignored her comment.

She had a daughter, Tina, who was deaf. I liked that name so much, that I named my daughter Tina.

New Orders

On May 24, 1969, just a little over two months after Larry was killed in Vietnam, I received orders to report to PorHue for three weeks combat training. I was then to be transferred to a new duty station in Vietnam. LTJG O'Donnell told me that since Larry had just died there, I had an option of not accepting, and that I could stay on the Kearsarge. Thinking about what my family had just been through,

I chose to stay on the Kearsarge. I could not put my parents through the agony of knowing I was in the middle of the action. The last thing my parents needed was to lose another son in that war. In my opinion, Larry had just repaid his debt for the wrecked car. LT.JG O'Donnell replied with this message:

> REQ to have EPDOPAC TC 142 cancelled in accordance with BUPERSINST 1300.28. Brother 1LT. Larry Maddy killed in RVN 3 Feb 1969. (US Army).

My Computer Connection

There was a young man from Minnesota in my department. His name was Lawrence Willey. Lawrence did not seem to have many friends. He was the nerd type of person. He looked as if he was only about fifteen years old. He was short and skinny. His hands were always dirty, and it looked as if he had not washed behind his ears for a month. I said something to another sailor about his being dirty, and he told me Lawrence was embarrassed to be naked in front of other men, so he chose not to bathe when anyone else was around. When we had water hours, you had to wait in a line to shower and he would not do it. I felt sorry for him.

He was a computer repair technician. Because I was friendly to Lawrence, he started hanging out with me. One day Lawrence asked me if I would like to see the computer room. Once again, I jumped on the opportunity to see and learn about something else.

The computer was a Univac and it was large enough to fill the room. I am not sure what it was used for, but they did keep track of the ship's supplies and inventories. Lawrence asked me if I wanted to play Black Jack on the computer. I was not sure if I should be playing games on this million-dollar computer, but he said it would be okay. I sat in front of a teletype keyboard and pressed the "N" or "Y" key when asked if I wanted another card. It was interesting. I played two games before we left.

Lawrence took me back to the computer room on another day and taught me how to use a key punch machine and how to use the card sorter. Later when I was organizing all the ship's test equipment, Lawrence suggested that we do that on the computer. With his help, I made an alphabetized list of every piece of test equipment on the ship.

This earned me a letter of recommendation from the officer in charge of OE division.

Petty officer Maddy has been involved in the setting up of an accounting system for 430 items of test equipment; and where it is, in what condition it is, and when it is due for calibration. He is quite capable of repair of nearly every piece of test equipment carried. He is recommended for re-enlistment.

There were over 400 pieces of equipment. It was interesting; every time someone was discharged, another Simpson volt/ohm meter would turn up missing. The ten new Simpson volt/ohm meters I bought in 1968 had all disappeared by 1969. All of those Snap-On tools I bought in Subic Bay were all missing, too.

Pot and Alcohol

One day I was taking inventory of test equipment. That equipment was all over the ship, at least in all the places with electronics. I went to a radar room to check their equipment. I had been there several other times, but this time the door was locked. I knocked on the door, but no one answered it. It was locked from the inside so I knew someone was in there. In the past the door was not locked, so I kept knocking. Finally I heard someone say who is it. I told them it was me and they said to wait a minute. It seemed as if it took a long time for them to answer the door, but they finally did. When I entered they asked whether I smelled anything? I said, "Yes it smells like smoke." I asked what they were smoking and they told me marijuana. They asked me if I wanted some. I declined. They asked me not to tell on them. I told them I didn't care what they did, if they didn't drag me into it.

I later thought, "Here we are in a war zone and these guys are high." Friends of mine told me later that many the guys in Nam were high on something. Some men had experienced such horrific things they could not get back to a normal life when they returned home. They stayed on drugs for the rest of their lives. This is the price our country pays to have freedom.

Some people would bring cases of CocaCola on board. After the ship was at sea for a couple of weeks, they would sell the cans for $1.00 each. Shoot, now we pay $1.50, but back then they normally sold for 10 cents. They were good entrepenuers. To cool their cans of Coke, they would spray them with the Co2 fire extinguisher. The man who checked the fire extinguishers thought they had a leaky extinguisher, but it was just being used to cool cans of Coke. It was a good thing we never had a fire in that compartment because the extinguisher was empty a lot of the time.

Some guys brought some rum to mix with that Coke. I didn't participate in doing those things because I didn't think it was worth the risk if we got caught.

CIA or FBI

There was a suspicion among the men that a CIA or FBI agent had been planted among us to find out if any of us were dealing in contraband or stealing supplies or leaking information. Lots of stuff was stolen off the ship on an ongoing basis, so it made sense there would be spies among us. When a new person of low rank who had no specific job and seemed to be very intelligent, he was pegged as a spy. No one would have anything to do with this person. One such man showed up in our division at the beginning of the cruise. No one had seen him before. I was told by other guys not to trust him or tell him anything. He would come into my office and just hang out. He never seemed to be around when you needed him for something.

One day when he was hanging out in my office, a couple of second class petty officers came into TGR and chained and padlocked him to a chair and told him they would throw him overboard if he did not reveal his true identity. The man never fought back and never told them yes or no. They left him there all day. It was quite funny. When they left the compartment, he was still chained to the chair and afterword I released him. No one will ever know the truth about that man. In my opinion, he was just a nerd who was an introvert that did not fit in. I never let him become my friend and I never told him anything. When the ship returned to Long Beach, he was not put on a paint chipping detail like the rest of us; he was transferred somewhere else.

Standing Watch

While we were in Sasebo, I was given watch duty in the aft port gun tub. The duty was simple. I was given a set of sound powered headphones and a nightstick. I had a four-hour watch. My job was to report anything that looked unusual or suspicious. My first thought was that we were in a peaceful port in Japan, what could possibly happen here. Then I thought, "What if our Vietnam enemy came to this port just to attack us in a place we would never expect it?" I took my job seriously and watched very attentively for the next four hours. As it turned out, nothing happened that was worthy of reporting. It was a boring job, but nothing was going to happen on my watch without it being reported. I was given this duty only once in my entire time in the Navy.

When I heard about the bombing of the USS Cole in Yemen in October 2000, my thoughts were how could this happen. Had the Navy changed its ways? I would have reported a boat or a swimmer when I was on watch. A swimmer (scuba diver) was one of the things I was told to be looking for. A boat would have been an easy thing to see and report. A year earlier, Lynn VanAcker had told me when he was on river patrol in Vietnam they were always on the lookout for swimmers. He told me they dropped concussion grenades over the side about every 15 minutes to kill swimmers who might attach an explosive device on the ship's hull.

A Journey in Japan

During my second trip to Vietnam, my mother sent me a newspaper clipping about LT. Gary Howlett being wounded in action in Vietnam. Gary and my bother Larry enlisted into the Army together. Gary and I attended high school together. Gary had been taken to an Army hospital in Zama, Japan. He was recovering from a leg wound.

Gary & Larry

On April 28, 1969, I decided I would go visit Gary when we entered Yokosuka. Up until now, all of my activities in foreign countries were with a group tour, or I was with a couple of my friends. This would be my first time going it alone. First, I bought a map of Japan at the base PX. I told the clerk, who spoke English, where I wanted to go and asked how to get there. He told me to take the train from Yokohama to Zama. It would be about an 80-mile trip. That did not sound difficult. The man at the ticket booth also spoke English and he basically, told me the same thing as he sold me a round trip ticket. There were many trains going to many destinations, so now where should I go? No one spoke English and the signs were all Japanese. I only knew two words of Japanese, sayonara (good-bye) and hayaku (faster). I went back to the ticket booth and asked which train I should take. He told me a train would stop at the platform number three. When the doors open, just get on. He said if I missed it, there would be another one in about 30 minutes.

I went over to platform three, stood about eight feet from the track, and waited. As the clock ticked and it got closer to the time the train would arrive, the crowd started to build. By the time the train arrived, there were hundreds of Japanese people standing on the platform. They were only about four to five feet tall. I was at least a foot taller and stood out like a giant. It was a weird feeling. When the train arrived, the doors opened, and those people started pushing. I was courteous and stepped aside to let them pass. The doors closed and the train sped away, and I was not on it. What the heck! I was being polite and was pushed away from the door and missed the train because of it. There was a man holding hands with a woman, and he made it on the train and she did not. The door closed on their hands.

When the next train arrived, I was not going to be left standing on the platform. The doors opened and I pushed all those little people out of my way and walked right onto the train. I was bigger and stronger than they were. They all looked at me as if I was a bully. They stared at me as if I did not belong there.

This was one of Japan's famous hundred miles per hour electric trains. There were about five stops on the way. That was my next problem; I did not know where to get off. The train went so fast I could not read any of the signs. There were limited seats and most of the people were standing, including me. The height of the windows, were designed for short people. I had to duck down to look out the window. That made it even harder to see signs outside the train. Imagine... I am too tall to see out the window, I cannot read Japanese, and the train is traveling 100 MPH, so I only have split seconds to look at a sign along the track. At each stop, people exited the train and the further we went the fewer people there were on the train. A young Japanese girl intuitively assumed I was probably going to the Army base. She did not speak English, but she motioned for me to get off at the right stop. I trusted her and got off the train. Now what? Where do I go from here? That same girl got off and pointed down the street, so I walked in that direction. Sure enough, there was an Army base about a mile away.

At the front gate, they directed me to the hospital room where Gary was staying. He was not there, and no one knew for sure where he was. I left a note on Gary's bunk letting him know I had been there.

Now getting back should be a lot easier because the train would end up at the Yokohama depot. Getting on the train at Zama was not the same because it was a rural stop and there were not hundreds of people pushing and shoving me. My whole trip lasted about 6 hours. This was an interesting experience. It just shows that you can communicate with hand gestures, smiles, and intuition.

Subic Bay

My first cruise in 1967-68 was through the winter months and it was quite cool. On the second cruise, it was during the summer months, and it was much hotter and more humid. Frequently the air conditioner was out or just could not keep up. Sleeping at night was often miserable. I would wake up with headaches more often than not. On top of the food not being that great, I often did not have an appetite because of the hot weather. Then with being on water hours and having no shower for days, things could get miserable. The Captain allowed sunbathing on the flight deck. He also allowed sleeping on the flight deck under the stars on hot nights if there were no air operations going on.

On May 17, 1969 we entered Subic Bay at 08:30. Liberty call went at 09:30, but the first and most important thing I had to do was get the test equipment sent off to calibration. It took about four hours to get this task completed, and I was saturated with sweat when I finished. The temperature was in the nineties.

There was a mail call at 13:00 and I received a letter from Lynn VanAcker and a package from home. The package included a birthday cake from mom. We frosted the cake and had a little party. The guys sang Happy Birthday to me. My actual birthday was 15 days earlier.

At 15:00 I finally went on liberty. I had very little appetite for Olongapo City, so this time I stayed on the base in Subic Bay. I went to the Navy Exchange and bought a Canon booster for my camera. In addition, I checked on the price of a new 1970 Mercury Cougar. There was an American car dealership is Subic Bay who was selling new cars. You could buy it there and pay no sales taxes, then pick it up in Detroit. You would skip the car dealership and save delivery charges. I purchased a new 1970 Mercury Cougar, and I was to pick it up when we returned to the states. I put every bell and whistle that was available at the time on it. I made a down payment and started making payments right away. If for some reason I could not follow through, I could cancel the deal and get my money back.

My dad had asked me if I could find him a statue of The Thinker. I told him I would look. In every port I shopped for a statue of The Thinker. There was a store on the base in Subic Bay which sold woodcarvings. I found this 14-inch high, rough-cut carving of The Thinker. It probably was not exactly what my dad had envisioned; it was going to be this or nothing. In all the ports we had been to, this was the only one I found, so I bought it. It really grows on you, and I like it. He placed it on the hearth by the fireplace and 20 years later, when he passed away, it was still there. I now have it on my hearth. As I said before, someday you'll probably get back the gifts you give your parents.

Roy Moruss, Bill Goss, John Hand, and I went for supper in the Spanish Gate and then to the Sky Club where we saw the Paulettes, a singing group that toured Vietnam entertaining the troops.

This started with an innocent trip to the Cubi Pt. E.M. Club. When we arrived, the waitress told us it was Veterans Day and the drinks were only ten cents each.

John Hand said, "Veterans Day is in November and this is May."

She said, "Here at the EM club it is Veterans Day every day and it is happy hour all the time."

At those prices, we were drinking Singapore slings about as fast as she could bring them to the table. A Singapore Sling is grenadine in the bottom of a Collins glass, filled with ice, then gin with equal parts of sweet and sour, cherry brandy, and soda, serve unstirred, and garnished with a cherry. You can see the results. We had been there for a while when we had a photo taken showing us to be a little silly.

AE2 Tripp, John Hand, waitress, Bill Goss, Me, Roy Moruss

We returned to the ship by 23:00. On the way back to the ship, Bill kept telling the taxi driver "hayaku." When we got back to the pier, Bill told the driver he was such a good driver he would give him a nice tip. It was not until the next morning that he realized he had given him a $20 tip for a $2 cab fare. That was a good night for the cabby!

That was also the night John Hand hit his head on the table in the berthing compartment. He was taken up to sickbay to get stitches. He said he remembers the corpsmen were discussing whether to use a baseball stitch or something else, but he did not remember much else after that. I guess John was pretty well self-anesthetized at that point.

While John was in sickbay, another sailor was there getting a shot for VD. John said the corpsman drew a target with an ink pen, on the sailors butt before they injected him with the shot. When they were finished, they said to him, “Well was it worth it?”

Water Skiing

On May 20, 1969 three other guys and I went over to a beach at Grande Island. The Navy had free cokes, beer, and hamburgers. The beer was warm, and I remember someone saying that is how it is served in Europe. There was a speedboat available for water skiing. After a few beers, I decided to water ski. I had done it before on the Illinois-Mississippi canal seven years earlier. The driver would take you from the beach into a large loop in the bay and bring you back to the beach. We started in water that was about three feet deep and returned to the same spot. Some of the experienced guys would let go of the rope on the final turn and return to the beach and step out of the skis into the sand. I had never done that, but thought I could. When the boat was coming around the final turn, I went out to the side to get maximum speed before letting go of the rope. I let go of the rope way too late, and when the skis hit the sand, they stopped, but I did not. I came out of the skis, flew about ten feet into the air, did three summersaults and landed in the sand, on my back about 30 feet away. It knocked the wind out of me. I laid there for a few minutes to catch my breath. The guys came running over to me to see if I was okay. I stood up, and not wanting to admit I screwed the pooch, said, “Let me see you guys top that one.” They looked at me and said, “No, you win!”

Broken Penis

While in Subic Bay, one of the guys in my division had intimate time with a girl in Olongapo City. I will not mention his name so not to embarrass him. I am not sure what went wrong, but the next day he was lying in his rack with one leg hanging over each side. He had a towel lying over his privates. When I saw him, I asked what is going on here. He told me he had broken his penis. Really, you can do that? His privates were swollen to the size of tennis balls and were black and blue. He said it did not hurt, but it looked awful. The ship’s doctor told him he would just have to wait until the swelling went down on its own. Word had soon spread around the ship and men were coming into our birthing compartment to see it. There was a steady line of men for an hour. I felt sorry for the poor guy. He was the last guy I would have guessed would end up in that predicament. A few days later, he was back to normal. I guess he had a blood vessel break and it caused the swelling. He must have had some wild time with her.

Day in Manila

On May 22, 1969 the ship steamed to Manila and met ships of other nations to participate in a SEATO (South East Asia Treaty Organization) training exercise, "Operation Sea Spirit.” There were sailors from United States, Philippines, Thailand, Australia, France, Pakistan, and New Zealand in Manila that day.

The Kearsarge had anchored in the harbor at Manila. Before the SEATO exercises started, we had a couple days liberty in Manila. I toured Rizal Park and The Rizal Monument. Rizal Monument is the equivalent of our tomb of the Unknown Soldier. There were many people flying kites in a kite-flying contest at the park.

After a day of touring the city, we all gathered back at the Manila Hilton Hotel. I was standing at an outdoor bar in a grassy area when I felt something inside my trouser leg. I looked down and saw a bulge on my thigh and it was moving. I thought a mouse had gone up my leg. I grabbed my pants in the spot and squeezed hard. Whatever it was, it started screeching. I let go, and it ran down the inside of my pant leg and out into the grass. It was a salamander lizard.

We went to a restaurant in the Hilton hotel to have dinner. A man of European descent was alone at a table next to us. He spoke with a broken English accent. He said he was a businessman. He offered to buy a round of drinks for us. Then he offered to have us sit with him for dinner; we did. After a lot of small talk, he started to ask about our jobs and the ship's movements. I started to tell him when we were leaving port and one of the guys told me to be quiet. We were taught in Boot camp, "Loose lips, sink ships." My first thought was "oh, not here, not this guy." Then I thought to myself, "Why not here, why not this man." That is how the KGB works. We all agreed he was probably KGB. We did not tell him anything other than name and rank. Our ship's movements were supposed to be classified information. However, it was strange that every port we entered the girls in the bars knew when we were arriving, how long we were going to be there, and when we were leaving. But, they never knew where we were going after leaving port. That was amazing because most of the time we did not know all those details.

Here is another story I had forgotten about until John Hand reminded me in an e-mail correspondence. We walked into a bar that was full of Australian sailors. They were yelling, laughing, and drinking lots of beer. We were curious as to what all the noise was about. So, they proceeded to show us. One of their men dropped his trousers and skivvies to his knees. He then took a rolled up newspaper and stuck it into his butt crack. They then proceeded to set the paper on fire. As he danced around on the stage, his buddies threw beer at him to put out the flames. They said, "It is the dance of the flaming asshole." Those Aussies were a crazy bunch, but they sure had fun.

They placed a bet with the American sailors that they could drink more beer. Guess what; the Australians won. A bet like that gets everyone drunk. I did not participate. I had already had that hangover experience on the train a couple of years earlier, and I had no plans of doing that again.

The Navy brought landing craft, "mike boats" to Manila to transport all navies to and from the ships anchored in the harbor. This landing craft was the same as you see when the marines landed on the beaches in WWII. Just after the sun went down, I decided to return to the Kearsarge. When we were going back to the ship, the waves were high and the craft was rocking and rolling a lot. Some of those men who had been drinking a lot started to vomit. They just took a bucket of water and washed down the deck. There were craft coming in and some were going out. The driver yelled for everyone to stay down so he could see where he was going. I said, "Yeah, we don't want to run into one of those craft coming in." Bang! Right then we hit an incoming craft. We hit them head on and everyone fell to the deck from the impact. We were wearing our white uniforms, and I was on my knees in the vomit. What a mess! No one was hurt, so the driver backed up and we continued our trip. The first stop we made was at the Australian carrier HMAS Melbourne. A drunken American sailor started to go up the ladder. Everyone yelled, "Hey buddy, you are getting off at the wrong ship." He turned around and yelled back, "Hey Mate, this is my bloody ship." It turns out he was an Australian sailor, who had swapped uniforms with an American sailor. When I finally arrived back on board the Kearsarge, I was soaking wet from water from the rough seas.

Disaster at Sea

During the SEATO exercise, one of our destroyers, the USS Frank E. Evans DD754, was operating with the Australian carrier, HMAS Melbourne. On the early morning of Jun 3, 1969, the Melbourne collided with the Evans, striking her dead amid ship and cut the destroyer in two. The forward half of the Evans sunk almost immediately with heavy loss of life. USS Kearsarge and her aircraft took part in the rescue operations. The survivors were brought aboard and were taken to Subic Bay.

The following is from a journal I kept on that day.

The Day 3 June 1969

I am not in a good mood nor am I very happy. The SEATO exercise "Sea Spirit" formed by seven countries making up a task group of 22 ships off the coast of Borneo, came to a smashing halt at 02:30 on 3 June 1969. The ships HMAS Melbourne and USS Frank E. Evans had a collision, which resulted in the Evans being cut in half just aft of the forward stacks. Being this time of the day, most of the crew of the Evans was sleeping. It has been said, by a few of the crewmembers, they were better off being asleep. If there had been more men awake and moving about there would have been more persons in the area of the cut. The Evans had a crew of 273 men. It has been reported that 196 were rescued. There has only been one man reported dead, the rest of the crew are missing at sea. The Melbourne had no fatalities. It has not been said as to whether they had any causality. The Melbourne was badly damaged in the bow and therefore was not capable of staying out at sea.

The bow of the Melbourne was designed for ramming. It is made of solid steel with a sharp wedge front and a reinforced backing. I have taken some pictures of both ships. Here are both the Melbourne and the remains of the Evans.

The forward section of the Evans sank ten and a half minutes after contact with the Melbourne. There is no possible way of recovering the Evans bow, the water is 7200 feet deep. So far, man is only capable of going 300 feet. The aft portion of the Evans is still afloat. It is taking on water but we are pumping it back out to take it in for salvage. The USS Larson DD-830, (a ship of equal size) is tied alongside the remains. It is not known for sure what reason there is for the collision.

The task force was in transient to Bangkok Thailand for a debriefing of the SEATO exercise. During this transient, the task group was going through tactical anti sub maneuvers. The Evans had been operating with Melbourne as a plane guard. By the way, Melbourne is a small carrier. Its flight deck is ten feet lower than that of the USS Kearsarge. The Melbourne had radioed the Evans twice, telling them that they were on a collision course. When the initial maneuvers had started, the ships were 1300 yards apart and somehow they managed to get together. The men of the Evans were tossed from there racks and were not warned in any way that a collision was eminent. Many of the men had jump over board. The waters were shark infested, but it is not known that they harmed anyone. There was a father and a son aboard, the father survived and the son was lost. In addition, three brothers aboard were lost.

At 04:00, I was awaked by public address announcement requesting for emergency blood donors. Again, at 04:30 the captain informed us of what had happened. I did not want to believe it was true; tears came to my eyes as I thought about how terrible it was. The Kearsarge was 43 miles away and it arrived at the scene by 05:00. It was at that time that I finally got up. I could no longer sleep. I went out on deck to see what had happened, but we had stopped about six miles away. We were within a seeing distance, but things were small on the horizon. Our planes and the Australian planes and boats searched the area for about 48 hours. After all known survivors had been rescued, we pulled in closer, and all persons were transferred from the Melbourne to the Kearsarge. Two of the survivors were put in OE division (My division).

We were not due back into Subic Bay until June 11; however, we are now going to be in on June 5, 1969. At 18:00, memorial services were held on the flight deck of the USS Kearsarge. We had a regular service with a 21-gun salute and taps. It reminded me so much of Larry's funeral from just a few months earlier. Only this time it was for 77 men. I thought of all of those mothers and fathers, wives, and girlfriends back home who suffer from the loss of a loved one. They all have my sympathy.

One of the survivors whom I talked with had just gotten out of boot camp and had only been aboard for 18 days. All of the survivors will be given a new issue of uniforms and $500 with thirty days leave and a choice of their next duty station. The thirty days leave will not be charged against them. Any personal items such as cameras will not be replaced.

First Man on the Moon

On June 20, 1969 when the moon landing happened, the Kearsarge was in the Gulf of Tonkin. I was in the berthing compartment watching it on the ship's closed circuit TV. Knowing it was a historic event, I took a black and white photo of the image on TV. It is not very high quality, but it is original.

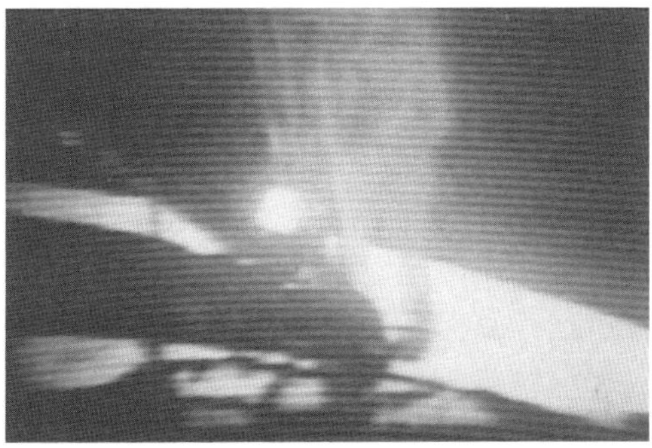

Rear Admiral Jerome H. King wanted to watch the Moon landing, but the TV signal was not coming in very well. ET's Allen Grant and John Hand were sent up on the mast to adjust the TV antenna so the signal being broadcast from Da Nang could be picked up by the Kearsarge. They had to do this while at sea with the ship moving forward while rocking and rolling. This made my climb up the mast like a walk in the park.

Mike Daniels was a radar technician in my division. About a week before the Astronauts returned to earth, he was transferred to the USS Hornet for temporary duty. On July 24, 1969 the space capsule splashed down south of Wake Island in the Pacific Ocean about 2,200 miles southwest of Hawaii.

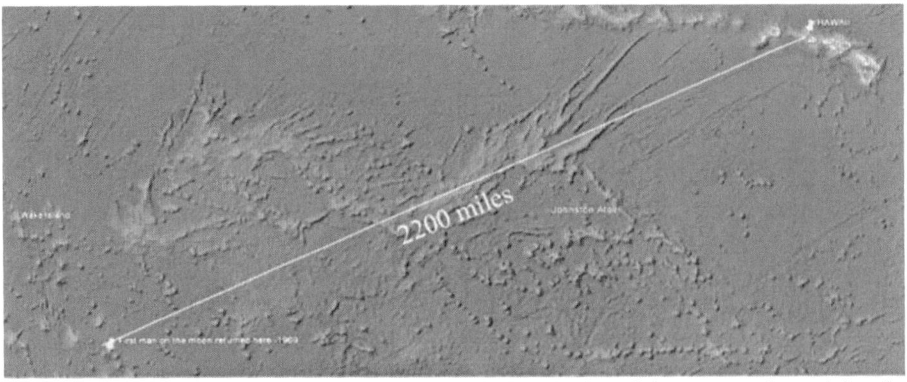

The Apollo 11 command module, Columbia, was placed in the hangar deck and roped off with a Marine guard in front of it. Mike could look at it from a distance, but not up close. He could see a foil coating that had blistered and ripped from the heat during re-entry. The next day Mike and another sailor were jogging in the hangar deck. When they were behind the capsule, they ducked under the rope, and ran up to the capsule, and grabbed a piece of the foil and ripped it off without missing a step in their stride.

When Mike returned to the Kearsarge, he showed me that piece of foil. It was about the size of a three by five inch note card. He said this is a piece of history. I told him I would give him $20 for it, but he turned me down. I upped the offer to $100 and he still turned me down. Mike went home with his piece of history.

Another Day

Later I found out we were operating among a group of small islands off South Vietnam, An Vinh and An Binh.

I worked toward getting custody cards signed by Fox Division officer. Fox Division runs the radar systems that control the five-inch guns. LCDR Bensinger was unhappy because I left the cards with Fox Division. I personally saw nothing wrong with it. I thought officers were supposed to be gentlemen and honest amongst themselves. I guess they are not, though.

The ship had a gun shoot and a rifle shoot one afternoon. We were within seeing distance of Hainan, a China owned island in the Gulf of Tonkin. The weather was beautiful.

July Fourth

On July 4 Captain Nearman shut down air operations and held a picnic for the crew on the flight deck. Many of the men put on shorts or swim suits, and hung out on the flight deck. They served hamburgers and hot dogs along with potato salad, baked beans, and potato chips. There were Vietnam surveillance photos on display for us to look at.

A bunch of us sat on the front of the flight deck with our legs hanging off the front edge. Today, the thought of doing that makes my knees weak. I took pictures of flying fish that were coming out of the water in front of the bow, flying about 10 to 20 feet, and then returning into the water. I had never seen anything like that before.

Some of the guys were shooting trap with shotguns. Others were playing volleyball in the hangar deck. It was a nice laidback day. The ship's band held a concert in the afternoon.

After the sun went down, they shot off signal flares for fireworks. They were all red, not the same as fireworks back home. I would like to have seen them shoot off the five-inch guns. Now that would have been real fire works with bombs bursting in air. It would have been the most expensive fireworks show ever.

Swim Call

While we were in the South China Sea, Captain Nearman stopped the ship and had a swim call. He said anyone who wants can go for a swim. He said, "And don't get any idea about swimming to land because that is two miles away, and that would be straight down." A swim call from a carrier was one of the photos in the recruiting brochure. I thought that if I ever get the opportunity, I want to do this. Now I had the chance.

They lowered a ladder with a floating platform at the water level by the quarterdeck. We dived in from there and swam about 300 feet alongside the ship. If you think the ship looks big when you are standing on the pier, you should see it from the water when you are swimming alongside it. On the starboard elevator, a rope net hung to the water. To get back on the ship we had to climb about 30 feet up the net. When I got to the top, I was exhausted.

As I stood on the elevator panting, someone said, "Are you crazy?"

I said, "It was not that bad."

He said, "No, this is the type of water where sharks are found."

I said, "That is why the Marines with rifles are in the boats."

He said, "I would trust the shark before he would trust the Marine. He would probably accidently shoot you instead."

If the Marines were as good with their rifles as I was with my BB gun, I had no reason to worry about it. Well, it is over now been there, done that! I always wanted to swim out in the open ocean in those exact conditions. It is one of those things I did while in the Navy; I will never be able to do it again. In my four years in the Navy this was the only swim call.

Purchases

Merchandise in Hong Kong and Japan was considerably cheaper than in the United States. I had purchased a camera and an onyx ring on my first cruise. It seemed as if everyone in the family wanted me to get them something during my second cruise. I had a whole laundry list of items to purchase. Things like birthstone rings, tape recorder, sweater, china, silverware, and a Thinker statue.

While I was in Japan, there was a china store on the base. There were so many choices it was overwhelming. I felt my mother would want something conservative and in the lower price range, so that is what I bought for her. Roy Moruss helped me carry the heavy boxes back to the ship.

All of the jewelry that everyone asked for was purchased in Hong Kong. The rings that I bought for mom and dad were both custom made. First, I picked out the stones and then a ring to set them on. A couple of days later I returned to pick them up. In all I bought about $1000 worth of merchandise for about $500. Great savings!

An Attempted Suicide

On July 9, 1969 we arrived in the port of Sasebo, Japan at 08:00. I had a regular workday until 15:30. I then went to town and spent time drinking beer. When I returned to the ship, I learned a man from the navigation department had jumped from the 06 level of the super structure and landed on the flight deck. That was about fifty feet high. As far as they know, he did this after it was dark. He was found about an hour after he jumped. He had two broken hips and a broken leg and his tongue was bitten in half. I never heard any more about the incident, but it was rumored that he had attempted to commit suicide.

Shore Patrol Duty

July 11, 1969 was a regular workday, and I had duty. It was also payday and I sent $65 home and loaned Roy Moruss $50.00.

I did very little work because I had 00:01 to 05:00 shore patrol duty. Yeah! Do you believe it: I had shore patrol duty. It was not long ago that I was making fun of the shore patrol. Now I was one. Moreover, it is in the same port. It was the first shore patrol I ever had and the only shore patrol duty during my time in the Navy. I was given an SP armband, an SP helmet, and a nightstick. We were sent off in groups of two.

In the wee hours of the morning we saw the ship's captain stagger aboard the ship, drunk. That was something I never thought I would witness.

Nothing happened on my patrol, but when I returned to the ship, I learned there were nine men mugged on the ship while I was gone. Later, it was determined that the crime had been committed by a couple of black men.

Racism

I was raised in a northern Illinois town that was all white people. I went to a school that was all white. I did not understand the black man's plight. My dad went on a business trip to Alabama when I was about seven years old. When he returned, he told me about how the blacks were treated in that state, things like sitting at the back of the bus, not being allowed to drink from a water fountain, etc. At that age, that information did not mean much to me because I could not relate to it. At about that same age I visited my Aunt Cleo's home in Waukegan, Illinois. She had a black woman housekeeper. We called the lady Aunt Jemima because she looked like the lady on the pancake box. I remember her being a very polite and friendly lady. She was the first black person I ever met. I had no racist feeling at all.

I do not understand how George Washington and Thomas Jefferson could declare America a free country and still have slaves. In my mind, they could have converted those people from slaves to employees. The difference between a slave and an employee is an employee is free to leave if he does not like how he is being treated by the boss; a slave is punished by the employer if he leaves. Based on that description all of the enlisted people in the Navy and Army were slaves. We were not there because we wanted to be, but because we were threatened with a punishment of prison if we did not go. In boot camp, we were told we could not use the toilet without a threat of punishment and that made me a slave, too. I have worked for people who treat their employees as if they own them. I was free to quit if I wanted to.

My first personal relationship with a black person was at my job at John Deere. I worked right beside two of them doing the same job, getting the same pay. I had no problem with them and they had none with me. I did not call them "nigger" and they did not call me "whitey" or "cracker." I did not understand what the big deal was and why they were being treated so terrible in the South. My next exposure to black people was in boot camp. There was not any reason to feel hatred toward them. They were just trainees like me.

In my division on the Kearsarge, there was a black man, First Class Petty Office George Ruff. George was a big man. He was a laid back, quiet kind of guy. He was friendly and easygoing and had a good sense of humor. I liked him. He did not talk Ebonics.

In the radio department, there was a nice looking black man who had an accent as if he was from a white neighborhood. He was friendly and had a great sense of humor. If you heard him on the phone, you would not know he was black from his voice. One day when he was hanging out with other black men, he was talking Ebonics and he sounded just like most blacks. Either he was pulling the wool over the black men's eyes, or he was doing it to the white man. My suspicion was that he wanted to fit in with the blacks and that what I knew of him was the real person.

I have never considered myself a racist, but I have developed a distrust of some blacks. What happened that caused me to become that way? There were a few things. One night when we were in Pearl Harbor, there was a huge fight on the pier where the Kearsarge was moored. It was black sailors attacking white sailors. I was not there and do not know what sparked it. It was disappointing because I thought we were all on the same side and that the Vietcong were our enemy.

The second thing was the destruction that the black people did during the 1965 Watts riots in California. Those people looted stores and burned buildings in their own neighborhood. That behavior does not exactly create an environment where I want to be near them.

The third incident happened while we were at sea. A few black men attacked and killed a couple of white men by stabbing them, with a screwdriver through their aorta. The black men were caught and sent to prison for murder. What was their motive? They hated "whitey." Now that is racism! This happened late at night. We were told not to be in the passageways at night to avoid this. It was not even safe on my own ship. It was the blacks, who created this unsafe environment. After 22:00 hours, the lights would be turned from white to red and the passage way would be darker. If you worked a night shift, it was necessary to be in the passageway. This incident made me more skeptical of blacks than anything else in my life did.

It is sad that because George Washington, Thomas Jefferson, James Madison, and other wealthy men would not free their slaves in 1776, we are still dealing with the aftermath two hundred years later. I am tired of hearing about the "poor me" thing from black people. I know many black people who have made good in America. They have great jobs, nice homes, and good families. Many of them have been more successful than white people I know. We have poor white people, too. Black people need to clean up their own act. It is not right to give a black preferential treatment because of the color of his skin. He must prove himself worthy just as a white person does. White people do not get ahead because they are white; they get ahead because they work hard and work smart. General Colin Powell said that as a black man it was not easy to rise among the ranks. It would not have been easy for me to get where he is either. It should not be easy for anyone to get to the top. The harder it is to get to the top, the more you will appreciate it when you get there. If it is hard, the better you will be as a leader when you get there. Too many people who get to the top without struggle are lousy leaders. You can see that in history all over the world. If a white man is picked on by someone like Chief I. M. Sheethead, he just says Sheethead is a bully, sucks it up, and lives with it until he can figure out how to deal with it. If a black man is picked on by Chief I. M. Sheethead, he says it is because he is black and calls it racism and wants new laws to protect him and wants compensation for the abuse. Is it any wonder that the whites have had enough of the race card BS?

If our government wants to get rid of racism, they need to quit asking people what race they are on all their forms. What difference does it make what race a person is? No college should lower its standards just so a minority can be admitted. If you do not meet the standard, you cannot get in, no matter what your skin color. On the other hand, no one should be rejected if qualified based on the standard set. If you are white and you hate black people, "Knock it off!" There are many reasons to dislike someone and the color of their skin is not one of them.

If you are black and you feel cheated, quit blaming the white man. Pull up your pants, learn to talk right, change your attitude, get an education, and fight through the system just like the rest of us. That is what investment broker Charles Payne did, and if he can do it, so can you. So did Dr. Ben Carson.

Hong Kong

On my second trip to Hong Kong, I did not take a tour to Kowloon; I spent more time in Hong Kong. The ship did not tie up at the pier, but instead it anchored in the middle of the harbor. We had to take a water shuttle to and from the ship. A trash barge was sent to the ship to pick up all our trash. I was told that because we had nuclear weapons on board, we were not allowed to tie at the pier.

I went to a tailor and had a custom suit made for me. The tailor made me a suit and six monogramed shirts. It fit perfectly and really looked sharp. Three other guys and I stayed the night at the Hong Kong Hilton. We put on our new suits and ate supper at the revolving restaurant on the top floor. We were living first class… for one night. That was the only time I wore that suit, as it was stolen from my locker in Long Beach within just a couple weeks after returning home. The tailor from whom we purchased our suits took all of us out for an authentic Chinese dinner. It was about a ten-course meal. If you have ever been to a Chinese restaurant and looked at the menu, we had a

taste of everything on the menu that night. The food was good.

During the next day, we went to the top of Queen Victoria Peak, and I took a picture of the Kearsarge sitting in the middle of the

Victoria Harbor. It was an awesome sight.

A bunch of us went into a Hong Kong bar just as the band went on break. The guys wanted to hear music, so they got on the bandstand and started playing the instruments themselves. The little Chinaman came from behind the bar and said, "You quit, me call shore patrol." The guys did not quit, he warned them one more time. Then he went behind the bar and picked up a phone. A couple of minutes later someone yelled, "Shore patrol is on its way." We all ran out the door just as the shore patrol was coming up the street. They walked right by us, because their destination was the bar. By the time they came back out, we were gone, laughing all the way down the street.

My friend Roy Moruss wanted to go to a Chinese specialty shop. We asked a Chinaman on the street where it was. He pointed north and said two blocks on the left hand corner. We walked up the street and stopped in front of the building we thought was the right place, but did not see a sign. We asked a ricksha driver where it was and he told us he would take us there. We jumped into his ricksha and away he went, down the street. He went one block and turned left. He went one more block and turned left. He went one more block and turned left again. He went one more block and said, "you are here." When he stopped, we were at the same corner where we started.

I said, "You just went around the block."

He said, "Yes, pwace you wooking for, on second fwoor, that be two dowwa" (about 30 cents U.S.)

We paid him the two dollars and thought what a con man. He probably thought, "I just took advantage of another stupid American." The rickshaw ride was worth the two dollars, and was the only one I ever took.

Destination CONUS

USS Kearsarge was leaving Hong Kong. When we cleared the harbor, the Captain opened his orders. We all learned the ship had been ordered back to Long Beach to begin deactivation. With the cruise having been cut short, the USS Kearsarge steamed directly from Hong Kong to San Diego where the air group disembarked. The ship left San Diego immediately and cruised to her home in Long Beach, arriving in the afternoon of September 4, 1969. Her final cruise was completed. CONUS means, Continental United States.

Top Secret Circuit Boards

Just as the Kearsarge was approaching the harbor of San Diego, California, the ship stopped. Now what? We have been on a six-month cruise that is just minutes from completion and we stop here!

Second Class Petty Officer Jones and a Seaman Denny were taking top-secret crypto circuit boards out of the radios and taking them to the vault for safekeeping. Seaman Denny had never been involved in this process before and misunderstood his instructions. He asked where they were taking them and one of the radioman jokingly said, "Were going to throw them over board." There were two special cases designed to hold all the boards. The cases were about the size of a two-foot cube. After removing enough boards to fill one of the cases, Second Class Petty Officer Jones told Seaman Denny to take the case down the passageway to a vault room and he would be right behind him when he finished filling the second case. Seaman Denny picked up the first case and left the radio room. A few minutes later, Petty Officer Jones followed with the second case. When Petty Officer Jones arrived at the vault, Seaman Denny was not there. Jones locked the second case in the vault and started looking for Denny. When he found Denny, Denny did not have the first case.

Jones asked Denny, "Where is the case?"

Denny said, "I threw it overboard like you told me to do."

Jones Said, "I told you to wait by the vault."

Jones quickly reported the incident to the Captain and the ship was immediately stopped. The officers started speculating as to whether the box would sink or float. A helicopter was launched immediately and they did not see any box floating, so it was assumed the box had sunk. Navy divers were brought from San Diego to look for the case. It took a while for the divers to arrive. We did not move until the case was found. They did not want those circuit boards to end up in the hands of another country, specifically Russia. This was a serious problem. Jones was really pissed that Denny was so stupid. I do not think Denny had a clearance to be handling those boards, so Jones was probably wrong for allowing him to help.

There were rumors in OE division that Jones and Denny would both be court marshaled for their incompetence. Jones was due to be discharged upon arrival in Long Beach, but was not released because of this incident. I do not know what happened to them because after we arrived in Long Beach, people were being reassigned to so many different places and jobs. I lost track of them.

Chapter 15 - My Last Four Months

Off Loading Ordinance

After being in Long Beach port for about five days, we went back out and anchored off Seal Beach, just south of Long Beach. While there, the entire ship's ordinance was off loaded to a barge and returned to the naval weapons station. This was a two-day process. All the ship's company was involved in passing ammo and other things hand by hand to the hangar deck. Some of the things we handled were highly explosive and dangerous. All safety precautions were taken. The Navy is good about that.

At 16:00 we were given liberty in Long Beach. The Navy provided a shuttle boat to get us from anchorage to a Long Beach Marina. We were told the last shuttle returning to the Kearsarge would be leaving at midnight. I arrived at the dock at 23:45, but could not return to the ship because of heavy fog. The petty officer in charge told us not to leave the dock because they would take us back to the ship as soon as the fog lifted. The fog did not lift until 08:00. About 50 sailors had been waiting all night. They were sleeping on the floor in the marina building. My back was hurting in the morning from sleeping on cold concrete all night. When the shuttle pulled up to the Kearsarge, a seal was sitting on the anchor chain. I guess that is why they call it Seal Beach.

Getting My New Car

On September 11, 1969 the ship was moved back to the dock in the Navy port of Long Beach. My dad thought he could get me a better deal on a new car from the Mercury dealer where he bought his car. I trusted his judgment and cancelled the car I ordered in the Philippines. On September 25 I took six days off and flew home to pick up my new 1969 Mercury Cougar. When I saw my new car, it did not have any of the bells and whistles I had ordered. The car was cheaper, but it was not the same. It was a left over from the 1969 models and was priced at $3,328.40. What I had ordered in the Philippines was a 1970 model. My dad told me I did not need all those things at my age and I could get

them later.

In those days, many of the young sailors were purchasing what are now called muscle cars and pony cars. There were many Mustangs and Camaros parked in the base parking lot. The pony cars were the Mercury Cougar, Ford Mustang, Chevy Camaro, Plymouth Barracuda, AMC Javelin, Dodge Challenger, and the Pontiac Firebird. The muscle cars were the AMC Machine, Buick Gran sport, Chevy Chevel SS, Dodge Charger, Ford Torino GT, Mercury Cyclone CJ, Oldsmobile 442, Plymouth Road Runner, and Pontiac GTO.

I stayed home two days then took four days to drive my new pony car back to Long Beach. I went the southern route through Missouri, Oklahoma, Texas, New Mexico, Arizona, and California. This was my first time in most of those states. The trip took two and half days.

A Date with Carol

After I returned with my new car, Bob Maehara called me and said his sister Carol was in from Hawaii and asked whether I would like to take her on a date. I said I would. Carol worked at a bank in Hawaii that was owned by a larger bank in Los Angeles. About once a year, she would come to LA for meetings. Carol and I were going to go to Disneyland, but I could not get tickets from the USO. We spent the day at the Hollywood Wax Museum and had pizza. She was a sweet girl and fun to be with, but she had to return to Hawaii for her job. I never saw Carol again after that.

Nancy

I was at a party in October when one of my shipmates introduced me to Nancy. She was from Canoga Park, California. Her parents had moved to the United States from Australia before she was born. She worked for the Pacific Bell phone company. I talked to her about my desire to go to Las Vegas, and she said we should go next weekend. Nancy, her cousin, Fran, and I went to Las Vegas for the weekend. Nancy was only 20 and was not allowed in the casinos. We stayed in the same motel room with the girls sleeping under the blanket and me sleeping on top of it. In those days, Nevada did not have a speed limit and when crossing the desert I drove my Cougar 120 miles per hour between Las Vegas and the California border.

Nancy's cousin Fran was the nanny for TV actor Ken Berry (F-Troop and Mama's Family). She took me to his house before leaving for Vegas; she needed to pick up her paycheck.

Nancy and I dated until I left the Navy in February 1970. We had only a few brief conversations on the phone after that. She started getting secretive about things and I decided to break it off with her.

Nancy's uncle worked at the Flamingo Hotel in Las Vegas as the head maître d'. He told me on Christmas Day of 1969 the price of gas was going up to $1 per gallon within the next year or so. I asked him how he could be so sure about such a thing. He told me oil tycoons had met in his hotel and decided they would create a gas shortage. So when people could not get gas, they would pay anything. I did not believe him, but he said, just to watch and see. By the end of 1971, it had doubled from 20 cents per gallon to 40 cents per gallon. By 1973 we had the gas shortage just as he predicted. It was 1988 when it reached $1.00 per gallon, which was 18 years later. He was right about the price and the gas shortage, but he was not right about when it would happen. The gas shortage was pre-planned by greedy oilmen. If I had not had knowledge about this a couple years in advance, I probably would not believe it.

The Decommission Process

The deactivation of the Kearsarge started immediately after ordinance was off loaded. The plan was to start in Long Beach, then move the ship to San Diego to finish the process. The Captain stood by his crew and took a firm stand against moving the ship out of her longtime homeport. He told the InActShips facility that the USS Kearsarge would deactivate in Long Beach. He would not move the dependents to San Diego. The ship could be towed to San Diego after the deactivation was complete and the crew was gone. I am glad he did that. I was able to stay in Long Beach where I was familiar with my surroundings.

Many of the ship's crew was transferred right away. They were sent to other ships or were sent home. A few men in my division were held back to help with deactivation, I was one of them. One of the first duties I performed was to clean the inside of the radio transmitters with tetracycline. Tetracycline was used because it washed away dirt with no residue and did not harm components. When a radio was clean, it was shrink wrapped in plastic to keep out dust and moisture. If the ship were needed in the future, they would only need to peel off the plastic and turn it back on. There were three radio rooms on the ship with as many as 20 radios in each, so it took a few weeks to complete this task.

I also worked in other areas. The one thing I never expected was to run a sand blaster and a pneumatic chipper. I was amazed at the huge chunks of metal rust that came off while running the chipper. I started to think the ship was a huge bucket of rust that had just been painted over for 20 years. When I saw that, I was glad it was being decommissioned and that I did not have to make another sea voyage with her. If it had been hit with a bomb, it probably would have just fallen apart. I may be exaggerating a little, but it was a lot worse than it looked on the surface.

In the stem of the ship, there was a deck space inside a passageway that was about 25 feet long by 4 feet wide. The chief had given another sailor (I do not remember his name) and I a five-gallon bucket of gray deck paint. He told us we could quit for the day after we sanded and painted that deck. The other sailor said to me, we would be done in about two minutes. I looked at him and said, “What!” At that point, he took the bucket and poured about two gallons of paint on the deck. He took a mop and swirled it around. The paint was about a half-inch thick on the deck. He then took the rest of the paint in the bucket and the mop and threw it over board into the harbor. I watched as the paint bubbled to the surface and created interesting shapes in the water. The sailor looked at me and said “We are done; let’s go.” What should have been an all day job was done in ten minutes. It was not done properly, however. The next day I looked at the deck and the paint had dried as a wrinkled mess. The chief looked at it and was not happy about it, but he did not ask us to redo it. We were re-assigned to the chipper duty in the hangar deck.

As different tasks were completed, more and more men were transferred away. Eventually the cooks were gone, and we were given passes to eat at the base commissary. As time passed, we were even transferred to a sleeping barge parked next to the Kearsarge. In November we moved into an apartment. The Navy paid us $175.00 for off-base housing and food. So, three other men and I moved to an apartment about 30 miles away near Paramount, California. Since I had a car and the other men did not, I was the official taxi. Therefore, I did not have to contribute to the rent. That gave me $105.00 for car expenses. Gas in California was 20 cents a gallon back then. We all kicked in money for food. Robert Clapper and Mike Daniels wanted to cook the meals, so McDonald and I agreed to do the dishes and keep the place clean.

Night Watch

In the final days, I was assigned to guard duty. There were only about 20 or so people left attached to the Kearsarge. There were only two people on security watch on the entire ship. Each watch was a four-hour shift then another two men would replace us. My shift was from midnight to 04:00. It was weird being on that huge ship with only one other person in the wee hours of the morning. I made the most of it, though, by going into places that were off limits before. Places like, the captain's quarters, the bridge, and the ship's hospital. I explored the ship as much as I could that night. I never saw that other person and wondered if there really was another person. The ship is so big that someone could have boarded it and taken whatever he wanted, and left and I would never have known it.

There were two bronze plaques on the quarterdeck commemorating the two space capsules recovered by the USS Kearsarge. I considered taking them home as a souvenir. It would have been easy to do since I was there late at night and no one else was around. Eventually, someone would have noticed it was gone, and the Navy would just look at the log to see who was on board at the time. That would have been me. I never would have been able to brag about having them because it would be a risk that someone would report me. I decided to leave them right where they were. I wonder what happened to them since the ship was scrapped.

Processing Out

In January 1970 the remaining men and I were asked where we wanted to go for our next duty. The personnel office told me I had two options, with the first being duty on the USS Kittyhawk. The Kittyhawk was stationed in San Francisco and was getting ready to depart for a tour in Vietnam. I wanted to see San Francisco, but I really did not want to make another tour to Vietnam. My time in the Navy was going to expire in about three months, in the middle of the cruise, and I did not want to fly back from the Far East. My second choice was to get an early discharge. I said, "Do I have stupid written on my forehead?' I took the early discharge. I could always go to San Francisco some other time.

Here is another story of a stupid military event. I was required to have a medical exam before my release from the Navy. During the exam, the Navy doctor told me they would have to keep me in the Navy because my blood pressure was above average. I said, "Are you serious?" First, they were not going to let me into the Navy because of high blood pressure, and now they were not going to let me out because of it. This just did not make any sense. When I told the doctor this story, he just signed the papers and let me go.

On February 11 the next thing for me to do was collect my final pay and all my Navy records. I went to the office to pick up my money and my records and was almost finished processing out.

There was one last thing I was required to do before I could go home. I needed to have a new Navy Reserve ID card issued. When you are on active duty, you carry a green card, and when you are inactive, you carry a pink card. I filed all the necessary paper work and sat for a new photo for the card. The petty officer who was taking the picture told me my mustache needed to be trimmed and that I should go home, trim it, and come back the next day. I complied and was back the next day. The petty officer told me it was not good enough and to trim it again and come back the next day. I had nowhere to go because I had moved out of my apartment, and the ship was decommissioned. I was heading for home immediately after my picture. I politely said, "Okay," with no argument and walked out of the building. When I got in my car, I thought, "What else do I need? I have checked out everywhere else." I had my final pay and all my papers, so why bother with that ID. I drove out the gate, flipped them the bird, and never looked back.

This is a picture of me from just a couple of days before I went in to have my picture taken. Do you think there is a problem? It was just more Navy BS.

So, It Was Done...

The USS Kearsarge was decommissioned in February 13, 1970 in Long Beach, California. The following three years it was in the Reserve Fleet, and (CVS-33) was stricken from the Naval Vessel Register in May 1, 1973. She was sold for scrapping in January 18, 1974.

A few years later I had talked to a man who had been on the Kearsarge and he told me it had been scrapped. At that moment I thought why did they spend all that time and money to put the ship into moth balls if they were just going to cut it up a couple years later? The paint that was applied so thick didn't matter after all. The de-humidifier at the bottom of the ocean does not matter because it would have been junked anyway. That's the military way!

Nearly 290,000 men have called the USS Kearsarge home, and she had 23 commanding officers during her 24-year life. I wonder what happened to the thousands of electronic technicians the Navy trained? Did they stay in that field, or have they gone back to their old civilian jobs, or something entirely different?

Now it is all just a memory of times past. The ship is gone, many of the people have passed, and the only thing we have left are the memories left behind by someone like me. The men stood for freedom and fought against communisum and socialism. Now the United States has more of both of those groups than many people think possible. People with that kind of thinking were considered the enemy. Now they are some of the politicians right here in America. They are the enemy. What a shame!

CVS-33 Travel Summary

I put together the ship's movements the best I could. There is a lot of conflicting information, so this may have some flaws. Most of the information I gathered was a match, so I am confident it is at least 99% accurate.

1945 – Built in New York Naval Yard
Apr 1946 – Stationed in Norfolk, Virginia
Jan 1947 to Feb 1947 – Guantanamo Bay, Cuba
Jun 1947 to Aug 1947 – Sailed to England, Scotland, and Sweden
Aug 1947 to Jun 1948 – East coast maneuvers
Jun 1948 to Oct 1948 – Mediterranean cruise
Jan 1950 to Feb 1950 – Bremerton, Washington
Jun 1950 to Feb 1952 – Decommissioned for stronger deck and new island
Mar 1952 to Aug 1952 – Long Beach, California – Sea trials and shakedown
Aug 1952 – Departed San Diego, California
Sep 1952 – Arrived in Yokosuka, Japan
Oct 1952 – Became CVA-33
Sep 1952 to Feb 1953 – Far East cruise - Korean War
Apr 1953 to Jan 1954 – Far East cruise – Formosa Straits
Jan 1954 to Oct 1954 – CONUS
1954 – "The Caine Mutiny" movie to depict Admiral Halsey's Flag-ship
Oct 1954 to May 1955 – Far East cruise – Tachen Islands
May 1955 to Jan 1957 – Hurricane bow and angled flight deck added
Jan 1957 – San Francisco, California
Jan 1958 – Became CVS-33
Sep 1959 to Mar 1960 – Far East cruise – Japan
Mar 1960 – CONUS
Oct 1960 – chases Russian sub to Bearing Sea
Mar 1961 to Oct 1961 – Far East cruise – Laos
Oct 1961 to Mar 1962 – Seattle, Washington – more modernization
Mar 1962 – crash at sea with Oriana
Aug 1962 to Oct 1962 – Hawaii - Sea trials and shakedown
Oct 1962 – Hawaii – retrieved astronaut Walter Schirra
May 1963 – Hawaii – retrieved astronaut Gordon Cooper
Jun 1963 to Dec 1963 – Far East cruise – Vietnam*
Jun 1964 to Dec 1964 – Far East cruise – Vietnam*
Apr 1965 to Dec 1965 – Far East cruise – Vietnam*
Jun 1966 to Jan 1967 – Far East cruise – Vietnam*
Jun 1967 to Feb 1968 – Far East cruise – Vietnam*
Feb 1968 to Mar 1968 – Far East cruise – Korea

Mar 1968 to Apr 1968 – Far East cruise – Vietnam*
Apr 1698 to 12 1968 – Long Beach, California – Dry dock
Jan 1969 to Mar 1969 – Long Beach, California – Sea trials
Mar 1969 – Hawaii – "Tora! Tora! Tora!" movie
Mar 1969 to Sep 1969 – Far East cruise – Vietnam*
Sep 1969 to Feb 1970 – Long Beach, California - decommissioned
Jan 1974 – Scrapped.

*These cruises she stopped in the the ports of Honk Kong, Subic Bay, Sasebo, and Yokosuka.

The Fourth Kearsarge 1992

Amphibious assault USS Kearsarge (LHD-3)

But wait! The name Kearsasrge lives on. USS Kearsarge (LHD-3) keel was laid down on February 6, 1990 at Litton-Ingalls Ship building Corporation of Pascagoula, Mississippi.

Ingalls built the USS Kearsarge using efficient pre-outfitting and modular construction techniques. Hundreds of smaller sub-assemblies containing piping, ventilation ducting and other hardware, as well as major machinery equipment, generators, and electrical panels, were constructed. The sub-assemblies were then joined with others to form assemblies, which were in turn welded together, to form five completed hull and superstructure modules. These giant modules, each weighing thousands of tons, were joined together on land to form the completed ship's hull. The result of this early outfitting was a ship that was over 70 percent complete at launch.[1]

She was launched on March 26, 1992 in a ceremony attended by then Chairman of the Joints Chiefs of Staff General Colin Powell. The warship was christened on May 16, 1992, and commissioned on October 16, 1993. This will be another story all its own.

Chapter 16 - Life after the Navy

No Fanfare

In February 1970 when I arrived home, there were no parades, no parties, no welcome home signs, and no shaking hands or thanks for a job well done. There was no post stress syndrome or exit interviews. It was simply that you are now home, so go get a job and get on with your life. I wasn't expecting anything, but when you compare it to how the WWII returning warriors were treated, this was definitely not the same. When the men went to war in 1941, they went by the train load. And when they came home, it was the same way. When we went to war for Vietnam, we left one at a time. And when we came home, it was the same way, one at a time. There was no big send off when we went, and there was no big reception when we returned.

To top it off the college kids were protesting the war on their campusses. I looked at them as a bunch of cowards who wanted the freedoms of America, but who were not willing to defend them.

The over all experience was a good one for me. I learned a lot during my time in the Navy. I learned about discipline, organization, cleanliness, and neatness. I learned about computers. I learned that some times you can get things done by not getting into all the details. I visited countries that I probably wouldn't have otherewise. I learned about other peoples' customs. I learned that people who look different and speak differently still have the same emotions and feelings as I do. I learned that people in this country who think of themselves as poor, have more than the average person in many countries. People in Mexico, China and the Philippines live with much less than the poorest in America.

There is no way that my 1960's $3.25 per hour job at John Deere would have afforded me six trips to Japan, two trips to Hong Kong, four trips to the Philippines, two trips to Hawaii, a trip to Mexico, and four trips to California.

According to Rear Admiral William Lee of the US Coast Guard, less than 8% of all US population has ever served in the military. That means I had experiences that 92% of Americans will never have. I rocked a 31,000 ton aircraft carrier. I swam in the South China Sea from a carrier. I sat on the front of a flight deck and took pictures of flying fish. I flew from a carrier.

As a member of the U.S. military, all my fellow soldiers and sailors and I are now in a category with George Washington, Abraham Lincoln, Dwight D. Eisenhower and John F. Kennedy as defenders of the Consitution of the United States of America. We are called veterans, and we are proud of our service and our country. We are an elite group of Americans.

Picking Up Where I Left Off

I returned to John Deere and received my old job back. They put me into the same position I had left. Some things had changed while I was gone. The pay had increased to about $4.50 per hour. My senoirity continued and I now had senoirity over many people who were hired after I left. When new positions opened I was able to bid into them over those people. They were not happy about it. To them I was a new kid on the block and needed to wait for my turn. The unions didn't see it that way. Within a few months, I had one of the best jobs, in the department. People waited years to get it.

Dee Dee

Another thing I did after getting home was to find Dee Dee. I was hoping I could pickup where that ended, too. She was still dating the guy she left me for. They were not married. She was working at a local cheese plant. She had chosen a life style that didn't fit into my future. I was disappointed! I decided if this was what she wanted, she wasn't the right person for me. I never saw her again.

Apollo 11 Spacecraft Foil

On September 10, 1970 I was at the Smithsonian Museum in Washington DC. They were in the process of building what is now known as the Air and Space Museum. Sitting along the wall without any security was the Apollo 11 command module, Columbia. There was still some of the foil on the sides. I walked up to it and ripped off a small piece, about the size of a silver dollar, and put it in my wallet. It was not as large as the piece Mike Daniels showed me. I carried that piece in my wallet for about 20 years. After moving four times, I have since misplaced it. In 1978 I was back in the new Smithsonian Air and Space Museum. It was completed and everything was on display, even the Apollo 11 command module. All of the foil was gone, and the capsule was enclosed in a clear plastic cocoon. You could not touch it anymore.

Old Friends

In 1971 I traveled back to California for a vacation with my new wife, Kathy, the mother of my daughter. While we were there, we stopped in Reseda, California to see Roy Moruss. Back in the summer of 1968 he had taken me there to meet his family. He told me that the actress Sally Fields used to live in the house behind him. Of course after she became a movie star, she moved away.

On our return trip to Illinois we stopped at Nogales, New Mexico and spent the night with Mike Lopez and his wife, Pam. The temperature was 110 degrees and now I wished I had not let my dad talk me out of all the bells and whistles on my new car. We had no airconditioning. It felt as if we were driving down the road with the heater on. Mike was working for the Annaconda Copper Mining Company. He took us on a tour of the area. It was nice to see him. I have not seen either of these men since. We sent Christmas cards for a while, but that dropped off, too. Life goes on with new friends in new places.

A Career Faded Away

I have often wondered where everyone ended up. In the civilian world there was not much of a call for our type of work, swapping out vaccum tubes. When I was ten, there was a repair shop for electrical appiances, radio, and TV's on Main Street in Granville IL. As the years passed, those places all went out of business. Today you can buy a toaster for $10. No one is going to pay $30 to repair a toaster they can replace for ten. No one repairs a radio or televsion anymore. They have become so reliable and so cheap and change in style so quickly, that people just buy a new one when the old one quits. Heck, they buy a new one even if the old one hasn't quit working, just so they have the lastest and greatest.

When I entered the Navy, a calculator was mechanical and sat on a desk. When I returned home they were electronic and fit in the palm of your hand and sold for $100. By 1985 they became the size of a credit card and were given away as promotional items. Today they are an app in our cellphone.

Almost everything electronic today is solid state technology, manufactured by robots in a far eastern country. Repairing a computer became swapping out circuit boards. Sixth graders do this today. There is no soldering parts in and out. There are no vaccuum tubes to replace.

I thought that maybe many of the men may have ended up in computers. Personally, I spent 35 years in the computer field as a consultant, operator, technician, or programmer, either on a full time or part time basis.

Navy Connection

About two years after I returned to John Deere, there was a job opening for a computer operator for a mainframe computer. Randy Larson, a co-worker, encouraged me to apply for the job. With no experience in the job I reluctantly applied. There were about 20 applicants for the job. After the interviews were completed, three others and I were in the final running. There was to be a test given to the four applicants. The test was exactly the same one I had taken at boot camp. At boot camp you were scored on what you got right out of what you answered and not what you got right of the over all test. This means if you only answer the questions you know, you can ace the test. That's what I did and I scored 100%. I beat out a high school teacher, a college grad, and the son of a John Deere executive. My Navy experience paid off.

During the final interview for the job I was taken into the computer room for a tour. During that tour the comptroller ask me if this looked like the computer I had worked with in the Navy. I never ran a computer in the Navy, but he didn't know that. I only played Blackjack on it. I said no, as this was an IBM and the Navy had a UNIVAC. No more questions were asked of me, and I didn't volunteer anything. As I said before, sometimes you can get things done by not getting into too many details. The experience I had with computers, thanks to Lawrence Willey's influence, helped me land this job. It also was helpful that the computer supervisor was an ex-Navy man, and he favored me over the other three men.

After I started my new job, I felt that maybe I was in over my head and told my new boss about my feelings. Glen Johnson, my new boss, just put his arm around my shoulder and said, "Don't worry, I am going to help you with everything you need to know". With that information I excelled at the job and became one of the top people in the department. I worked so hard at it that I soon knew more about the job than anyone in the department, and that included Glen. He pushed for me to get this job because I was an ex-Navy man. He did more for me in my life's career than anyone. I want to thank him here and now for his support. The experience I received with the influence of Glen Johnson helped me to land a six figure income job later in my life. It all happened because I befriended a little, nerdy, computer technician in the Navy.

My Family's Military Service

My brother Wayne entered the Air Force on November 1, 1964 and was discharged in January1969. He was stationed in Oklahoma, Argentina, and Panama. Wayne received an honorable discharge in 1970.

On August 30, 1966 I went into the Navy. I served on the USS Kearsarge from June 26, 1967 to February 1970, making two tours to Vietnam. On May 16, 1972 I received my honorable discharge from the Navy. This meant I was officially finished with the military.

My brother Larry enlisted in the Army on November 1, 1966 and was killed in action in Vietnam on February 3, 1969.

In September 1970, my brother John was drafted into the Army and was sent to Ethiopia. He received his honorable discharge in 1976.

In 1974 after the Vienam conflict was over, my bother Paul enlisted in the Marine Corp. and was shipped to Okinawa. With his enlistment, my family had now served in the Army, Navy, Air Force, and Marine Corp. Paul received an honorable discharge on January 10, 1980.

As the Billy Ray Cyrus song says, "Many gave some, and some gave all." While standing up for the American flag some had to fall. When you think about all your liberties, think about this, shamefully some gave nothing at all.

A Visit With Soldiers

In the early 70's a few years after I was home from the Navy, the soldier who was with Larry the day he was killed in Vietnam came to visit my parents. The story he told was eery to me. He said they were on a reconnaisance mission looking for the VietCong. They had not been successful at finding anyone. They reentered the chopper and started to leave when Larry spotted someone moving below them near a bush. They landed again to investigate. Larry took the lead and ran to the bush. Lying on his back was the enemy pointing a gun at Larry. He shot and Larry dropped and was dead immediately. The men on his team shot and killed the enemy.

This story sounds so much like the re-ocurring nightmare I had back in 1969. The similarities are down right eery. The fact that both were a one on one, and both required someone to shoot first, and the fact that Larry and I talked about this before he went there.

Giving Back

When I was in the Navy the USO had a facility in every town near a Navy base. They offered free sandwiches and soft drinks. They also had discounted or free tickets to movie theaters and amusement parks. I went to Disneyland several times for free with tickets the USO provided.

While we were in Vietnam, the USO show boarded the Kearsarge. The main attraction was Connie Francis, with her hit song, "Where the Boys are." Bob Hope had a big event every year for many years. There is a great deal of expense in taking all the celebrities on those tours. I really appreciated everything the USO did for me so it was one charity that I gave to for forty years.

Another charity that I felt was worth while is the DAV Disabled American Veterans. Those men gave a lot, and the least I could do was give back to them. I also recommend supporting Wounded Warriors Project.

Fireworks

On July 4 1978, I was shooting off fireworks in my back yard. I stepped into the house for a moment to get some more matches. When I returned, there was a city policeman standing on the patio. He asked me if I was the one who was shooting off all the fireworks.

I told him "No, it's not me." Right in front of me the picnic table was heaped full of firecrackers, bottle rockets and skyrockets.

He said, "You know here in Iowa, it's illegal to shoot fireworks."

I just looked him square in the eye and with a voice of authority said, "Did you serve in Vietnam?"

He said, "No."

I said, "I did, now get the hell off my property and let me celebrate America's freedom."

Without saying a word, he turned around and left and never came back. I continued to shoot my fireworks.

POW LT. Cdr. Charlie Plumb

On March 30, 1982 I met LT Cdr Charlie Plumb in Milwaukee, Wisconsin. Charlie was a POW from the Vietnam war. Charlie was a pilot who was stationed on the USS Kittyhawk. At 10:30 on the morning of May 16, 1967 Charlie was launched from the Kittyhawk to make a strike over Hanoi. He never came back. He was shot down and taken prisoner. On February 17, 1973, Charlie was released. He spent six years as a prisoner.[1] He wrote a book called "I'm No Hero". Everyone should read it. I spent about 30 minutes talking to him about our time in Vietnam and aboard a carrier. We were both in Vietnam during that time period, doing different jobs.

Charlie's wife divorced him while he was held captive. He had a hard time getting on with his life when he returned. Six years of his life just disappeared into oblivion. I want Charlie to know he has been in my thoughts since the day I met him.

The POW's and MIA's are real people with names and faces. Americans should get to know these people. They are my heroes.

What happened to the HMAS Melbourne?

HMAS Melbourne was sold in February 1985 to the China United Shipbuilding Company for $1.4 million, with the intention that she be towed to China and broken up for scrap. Prior to the ship's departure for China, the Royal Australian Navy stripped Melbourne of all electronic equipment and weapons, and welded her rudders into a fixed position so that she could not be reactivated. However, her steam catapult, arresting equipment, and mirror landing system were not removed. At that time, few western experts expected that the Chinese government would attempt to develop aircraft carriers in the future. The carrier departed Sydney on April 27, 1985, heading for Guangzhou, China under the tow of tug *De Ping* and arrived in China on June 13. The Australian government received a telex on this day, ing:

> *Please be advised that HMAS Melbourne arrived at Port Huangpu, intact and safely afloat, proud and majestic. She has been innocent, never once bowed to the natural or human force, in spite of the heavy storm and the talked about jinx.*

The ship was not scrapped immediately; instead, she was studied by Chinese naval architects and engineers as part of the nation's top-secret carrier development program. However, it is unclear whether the People's Liberation Army Navy (PLAN) orchestrated the acquisition of Melbourne or simply took advantage of the situation. Rear Admiral Zhang Zhaozhong, who is both the son of PLAN founder Zhang Aiping and a staff member at the National Defense College, has stated that the Navy was unaware of the purchase until Melbourne first arrived at Guangzhou. Melbourne was the largest warship any of the Chinese experts had seen, and they were surprised by the amount of equipment that was still in place. The PLAN subsequently arranged for the ship's flight deck and all the equipment associated with flying operations to be removed so that they could be studied in depth. Reports have circulated that either a replica of the flight deck, or the deck itself, was used for clandestine training of People's Liberation Army Navy pilots in carrier flight operations. It has also been claimed that the Royal Australian Navy received and "politely rejected" a request from the PLAN for blueprints of the ship's steam catapult. The carrier was not dismantled for many years. According to some rumors, she was not completely broken up until 2002. A 2012 article in *Jane's Navy International* stated that the large quantity of equipment recovered from *Melbourne* "undoubtedly helped" Admiral Liu Huaqing secure the Chinese government's support for his proposal to initiate a program to develop aircraft carriers for the Chinese Navy.[2]

If the United States ever ends up in a carrier war with China, we can blame the Australians for it.

Flags Of Honor

In 1992 my sister Clair Cook called me to tell me she had purchased a Memorial flag to Honor each of her brothers who served in the military, Wayne, Larry, Dale, John, and Paul. The flags are flown in the East park in Sheffield, Illinois during Memorial Day weekend. This was the first recognition I had received for my service in the Navy- twenty years after my discharge. In retrospect I am proud to be a Veteran of the United States Military. Among my favorite holidays are Memorial Day, July Fourth, and Veterans' Day, the days we fly the American flag.

Back In Hawaii

In the mid 1970's, my mom and dad visited Hawaii. They called Bob Maehara and he took them out for dinner one evening. They liked Bob; he was a Hawaiian ambassador to them, too.

In March 1995, I was back in Hawaii for the first time since my Navy days. I stayed with friends in Kauai for a few days and spent the remaining days in Maui. When I was at the Honolulu airport, I called the home of Bob Maehara. His brother answered. They were living in the same home. I asked if he remembered me; he said he did. I had sent Christmas cards to his family for years. Bob had told me that if I ever came back to Hawaii I should call him; he would show my family around the island. He said, "I do not care if you have ten kids, it will all be on me." That was quite an offer. I was not married at the time and was there with a friend.

I asked about his sister, Carol, and he said she was married and doing well. I asked if Bob was available to get together and he said no, that Bob had died a week ago. He died from hepatitis that he got from one of his patients. Bob had never married. He told me his mother had died a few years earlier. I wanted to ask Bob why he had such a short stay in the Army. I will never know. Carol died Dec 23, 2013.

In 2012, my daughter and I were in Hawaii and I called the house one more time only to find the number had been disconnected. She and I researched on the internet to find him, but had no luck. We did find information about Bob's death, but nothing about his sister or brother. I told my daughter it was all just a memory now and the connections are gone. That is how life is.

While I was in Hawaii, I finally did go to the USS Arizona Memorial. Everything you ever wanted to know about the war with Japan can be found there. I highly recommend it.

Commander Bucher

In October 2000 on a return trip from Utah, I stopped at Father Flanagan's Boys Town in Nebraska. I have contributed to Boys Town and wanted to see what they were doing there. I was introduced to a couple of young students with tattoos on their arms and necks. These boys were better behaved and showed more respect than kids show in my own neighborhood. If you ever wondered about Boys Town, I can tell you it has done many great things for many young boys. I was impressed with what I saw.

In the museum they featured students who graduated from Boys Town High School. One room was a hall of fame of boys who went on to do great things in life. Among them was a picture of Cmdr. Lloyd Mark "Pete" Bucher, Commanding Officer of the USS Pueblo. Bucher died - January 28, 2004.

Vietnam Wall

On July 23, 2001 my daughter, Tina and I visited the Vietnam Wall in Washington D.C. It was my first and, most likely, my last time there. This was a difficult place to be. I wanted to see it, but the emotions it brings out are so hard to endure. Many of my friends had been there; some of them had rubbed Larry's name with a pencil and sent it to me. We stopped at the Lincoln Memorial first. My emotions were building, just thinking about where we were headed next. The whole thing is overwhelming. We looked at the directory, which was organized by state. It told us which panel to look for to find Larry's name. The names are displayed based on date of death. When we found Larry's name on the panel 33 W, I could not hold back the tears. So many thoughts go through your head at this time. All the things that were, and all the things that could have been, all the things that never will be. I felt bad that my parents never had a chance to see it. Maybe it was better they did not. After I settled down, I put Larry's Bronze Star and Purple Heart next to his name and photographed the wall.

San Francisco

In October 2006 I finally took a vacation with my second wife, Debbie, to San Francisco, California. It was much better than it would have been back in 1970, and I didn't need to see the USS Kittyhawk. It was Fleet Week while we were there; the Blue Angels performed over the Bay by Fishermans Wharf. Times have changed; the people in San Francisco want the military out of their city, or so says our City Tour Guide. That is being hypocritcal because they want the freedoms afforded them by the military, but they don't want the connection to it. San Francisco and the U.S. Navy have a long history together.

Code Talkers

In 2007 I visited the Code Talkers museum in Cortez, Colorado. I was looking at all the native American men who had service in WWII as a code talkers. A native American woman asked if I had served. I told her that I had, and she thanked me for my service. That was the first time in 37 years that anyone outside of my family had thanked me for my military service. It brought tears to my eyes when I thought about a native American being grateful for my defending America when in the past it was the white man who overran the Indians.

While on that same trip we stopped at Taos Pueblo Indian Reservation at Taos, New Mexico. I talked with a young Indian woman who had inherited her

grandmother's home on the reservation. She said the homes have been passed from generation to generation for 800 years and now it was hers. The Pueblo Indians had lived a simple life that primarily revolved around survival, food gathering, and food preparation. I told her it was sad how the Europeans overran the American Indians. She replied that she was glad they had. She said she preferred to have all the technology that they have brought with them and created since. She said life today is much easier than the one her ansestors lived.

Meeting LT. Jim Brown

In July 2009, I was in Massillon, Ohio visiting my brother Wayne. Wayne had met a man from Canton who was a pilot on the USS Kearsarge in 1967 and 1968. Wayne made arrangements for me to meet LT. Jim Brown for breakfast. Jim was in VAW-111 Squadron. VAW-111, based at NAS North Island, in San Diego. They provided airborne early warning to all west coast based carriers utilizing detachments. Each detachment was composed of Grumman Tracer, E-1b aircraft. We called the plane the "Willy Fudd."

This was great for me as I never talked with any of the pilots while on board. Enlisted men were not allowed to be in the part of the ship where the officers worked and slept. I mailed Jim some footage of super eight movies I had taken and Jim mailed me this photo of his plane on a landing aproach. This is a view that very few people would ever have an opportunity to photograph. Navy pilots are a unique group of men and women. There are pilots and then there are carrier pilots. Carrier pilots are couragous people. They land a 150 MPH airplane on a moving postage stamp, and do it over and over.

Greeting Soldiers

In November 2009 I was at the Honolulu airport when soldiers were returning from Iraq. As they exited the airplane, my sister Alice and I stood by the gate and welcomed home each one of them with a hand shake and thanked them for their service. This is the least a person can do for someone who is risking their life so we can live ours in peace. It was important to me to do this, and they appreciated it.

Vanity Plates

In 2010 I purchased Vietnam Veteran vanity plates for my vehicle. It reads "KSARGE" and displays the Vietnam Service Ribbon. Someone already had "CVS-33." Since I have done this, many people have saluted me, shaken my hand and thanked me for my service to America. One day when I was driving on the interstate highway, a car pulled up next to me and saluted me. I wondered why he did that, then I remembered I had those plates on my car. He was thanking me for my time in the military. What a nice feeling that was.

In September 2010 I drove to within three miles of Mount Kearsarge in Sutton, New Hampshire while displaying my KSARGE plates. In my research I found the Kearsarge Inn located in North Conway, New Hampshire. I do plan to spend the night there one day.

WW II Survivor

At 2:00 P.M. on June 28, 2011, I watched a National Geographic special "The Lost Fleet of Guadalcanal." It is about the battle between the Japanese fleet and the US Navy at Guadalcanal. Between the Japanese and the USA, fifty ships were sunk. The story featured the USS Quincy (CA-39), a Heavy Cruiser. Robert Ballard,

who located the Titanic, was sent on a mission by the US Navy to find the USS Quincy.

During the early hours of Aug 9, 1942 while on patrol off Savo Island, just 14 miles north of Guadalcanal, her force was surprise attacked by Admiral Gunichi Mikawa's force. When the attacking Japanese cruisers trained their searchlights on Quincy, Quincy's guns were still trained forward. She was caught between two columns of Japanese cruisers. When they opened fire, she received a horrible pounding. "We're going down between them - give them hell!" said Captain Moore of the Quincy. She fought valiantly, but was not able to overcome the heavy firing from the Japanese cruisers. She sank at about 02:35 A.M. that day. She became one of the first ships that eventually nicknamed that area of the water "Iron Bottom Sound." Three hundred seventy lives were lost aboard the USS Quincy during the Battle of Savo Island, and there were 167 wounded. Eight hundred men were aboard the USS Quincy at the time.[3]

As I watched this story, I could not keep from shedding tears for the families and the men who gave their lives that day in August 1942.

Someone on the Japanese ship took this photo, just as the Quincy was burning and before it sank.

In the NatGeo story, they featured survivor Leonard Joslin, a petty officer signalman who was 25 years old at the time.

Leonard said that over the years he has had nightmares about that night in his life. He said that he dreams that the men are waving at him and they are alive again, and then they just fade away.

Now fast forward, five days to July 3, 2011

My daughter Tina, my son-in-law, Rob, and I went to the Wapsipinicon State Park in Anamosa, Iowa to visit the Ice Cave. Tina and Rob were going into the cave when an older man pulled up and said it was not that far into the end of the cave.

I said to him, "You have probably been going into that cave since you were a little boy,"

He said, "Yes."

I asked him if he had lived here all his life and he said that he had, except for the six years he was in the Navy. I told him I had been in the Navy. I asked him which ship he was stationed on. It was the USS Quincy. He said that it had sunk at Guadalcanal. I said to him that just five days ago I had watched the story about it on Netflix.

Leonard said, "Yes, I am in that story."

I told him... **You are the True American Hero**

Here I am with Leonard on July 3, 2011. Leonard is now 94 years old.

I asked Leonard for an autograph.

Leonard said, "It was about 01:30 and it was as dark as it could be, without a moon or any stars in the sky." He said, "The Japanese shined their spot lights on the Quincy and started firing on us immediately. He told me that there were 14 signalmen aboard the Quincy. He said that when the attack started, he and seven other signalmen were standing on the signal bridge. After the first explosion, the other six were all dead. He was hit with something himself and was stunned and bleeding. The first thing he did was reach for his arms and legs to see if he still had them. Then he ran to the other side of the ship to protect himself from further blasts. Shortly after that, the ship was taking on water, and it started to sink. So, they abandon ship. He was one of the lucky ones.

What are the odds of me running into Leonard Joslin just five days after watching this video?

All I could think is how lucky I was to survive war, as those ships do sink and that could have been me. During the Vietnam War, our ships were not attacked like that. Moreover, that was because of the work the sailors on the USS Kearsarge were doing.

Two years later on July 22, 2013 while staying at Rob and Tina's home, I decided that since I was so close to Leonard Joslin's home, I would stop by and visit him. I thought that maybe I should call first to give him a heads up that I was coming, but no one answered the phone. I thought that since he is 96 years old maybe he was not able to get to the phone, so I decided to stop anyway. When I arrived, his daughter told me he was at McDonald's, having coffee with his friends and that if I hurried I could catch them. I quickly drove to McDonald's.

When I entered McDonald's I did not see Leonard, but there were five older gentlemen sitting together. I knew that since I was in Iowa these people would be friendly, so I approached them and asked if they knew Leonard. They told me he just left and went back home. I showed them the book that I am writing and told them there was a story about Leonard in it. They started to tell me about Leonard, and I told them I knew that stuff. I wanted to show him what I had written and ask him some more questions. Two of the men were Korean War Veterans, and one was a Vietnam Veteran. One was physically unfit for duty, and the other never served. I told him, "Shame on you." The three who served said they thought that every person should be required to serve for two years after their nineteenth birthday. Those men felt that every man and woman who is physically and mentally fit should be required to serve six years of service to their country. Two years in active service and four as inactive. When inactive, they could be called back to active status with a 30 day notice. A person could choose from the Coast Guard, Navy, Air Force, Army, Marine Corp, or the Peace Corp. It was their opinion that no one should be allowed to be President of the United States and Commander in Chief unless he or she has served in one of the five branches of the military. How can you command people to war when you have no idea what you are doing to those people? How can you make good strategic decisions when you have no training in doing that? A man without military experience can never measure up to a John Kennedy no matter how smart he thinks he is. I said, "Yes, I agree with you."

I visited with these men for about 15 minutes then went back to Leonard's home.

When I arrived at Leonard's, he was working in his yard. I showed him the book I was writing and asked if he remembered meeting me two years ago. He said, "Yes." I told him I had met his friends and he said they were all over the age of 80, and they were retired schoolteachers except for one who was a farmer.

Leonard invited me into his home and we visited for three hours. While I was there, his phone rang and it took him a few minutes to get to it. He told me sometimes it stops ringing before he can get it answered. That answered my question about him not answering when I called. I had many questions I wanted to ask him. He and I both served in the Navy, 27 years apart. We both experienced crisis. The first thing I showed him was what I had written in my book about him. I told him that I saw him as a hero. He just smiled when I told him that. I then read to him what I had written about the National Defense Service Medal in both chapters two and sixteen and asked him if he thought my thoughts were valid. He said that they were.

I asked Leonard to tell me about boot camp. He said that he went to Great Lakes, the same as I did. When Leonard was there, he was involved in a boxing tournament. He had participated in Golden Gloves when in high school, but said he was not very good at it. He said the man they paired him with was taught how to box and was good at it. Leonard decided he would just hit the fancy boxer with all he had. In the first round, he hit his opponent and knocked him to the ground. In the second round, he decided he would knock him out. When his opponent came in for a hit, Leonard hit him square in the jaw. The man's knees buckled, and he fell to the ground. The referee got the man back on his feet and told them to continue the fight. Leonard said, "I am not going to hit him again." Just then, the man fell to the ground again and they declared Leonard the winner.

Leonard enlisted in the Navy in 1939 and asked for the USS Quincy because his older brother was on it and they could be together. They operated in the Atlantic dealing with German submarines. The ships worked out a code system to avoid being hit by a German sub;

they flew a black flag when a sub was spotted. When the black flag was raised by any of our ships, it meant Germans were present. When all ships were flying the black flag, they turned left and the Germans would miss our ships. The plan worked perfectly. At first, the Quincy did not have a black flag, so the chief asked Joslin and another sailor to make one. When they did not get it done right away, the chief wanted to know why. The other sailor said it was because Joslin did not know how to run a sewing machine.

Leonard's brother had been discharged from the Navy in 1941 before the war started, but was called back after Pearl Harbor was bombed. Leonard told him not to get back on the Quincy because it would be too dangerous for both of them to be on the same ship. That turned out to be a wise decision.

Leonard said one of his friends, a Cherokee Indian, named Martin, had studied airplanes, and knew every shape and size of the Japanese fleet. One day he and Martin observed a squadron of planes approaching. Leonard made a comment about how nice it looked in the sky with all those planes in formation. As they got closer Martin said, "Those are not ours, they are Japs." So they ran for cover.

I asked Leonard again about the night his ship was hit. He said the Japanese were coming from the north and the U.S. and Australian ships were coming from the south. There was no moon or stars in the sky, so it was pitch black. When the Japanese trained their spotlights on the USS Quincy, he saw three flashes of light between the spotlights. He knew that meant the Japanese had just fired their cannons.

Leonard yelled, "Hit the deck."

An officer saw them all lying on the deck and said, "That's not what sailors do; we are men, not mice, we face the enemy."

The officer stepped over the men on the deck, walked over to the handrail, and looked out across the ocean. A split second later, the projectile exploded and the officer's head was blown off. Six of the signalmen that were next to Leonard died too. Leonard managed to survive that blast. He was injured with shrapnel, but was still alive.

Not long after they were told to abandon ship. Martin told Leonard he could not take it anymore and jumped into the water from the signal bridge from about 50 feet up. When he hit the water, another explosion blew him out of the water and about the same height again. He felt like every bone in his body was broken, but after the rescue, to his amazement, the doctors found nothing was broken, and he was fine. As they were leaving the ship, Leonard's friend saw the bugle boy lying on the main deck. He ran over to him and said we have to abandon ship. Upon further inspection, he realized the poor kid was mangled so badly he would not be able to get up.

He told the bugle boy, "There is nothing I can do to help you."

The young boy said, "Tell my mother I love her," and then he died and his body went down with the ship.

After they were in the water, Leonard found that his life jacket was so badly damaged that it was worthless, so he took it off. The chief petty officer and another man who had lifejackets put him between them so he could float with them. A little later, they spotted a net that was balled up in its storage position. The chief had a penknife in his pocket; they used it to cut the tie strings. They all climbed into the net and floated inside it.

To top it off, after all the chaos, it started to rain and there was a lot of lighting. The flashes of lightning lit up the surface of the sea, and they could see ships in the area. The problem was that they did not know whose ships they were. They thought it was the Japanese.

The chief said, "We could be rescued by the Japs."

Leonard said, "I am not going to be picked up by the Japs."

When the lightning flashed, he could see Savo Island, which was about four miles away. He was determined to swim there. One of the men had brought a signal flashlight with him so they flashed a signal to the ship. If it were one of ours, they would know the code and would answer with a positive response. It was one of ours and the men were rescued.

The U.S. had a complete fleet of ships in and around Guadalcanal at the time. There were carriers, battleships, cruisers, and destroyers. Because of that, Leonard and the men thought they would be okay when this all started. What they did not know at that time was the carriers and battleships had left the area and they were not there to help. Three American cruisers were sunk that night. The Japanese also knew the battleships and the carriers were in the area, and after their attack, they left too. The Japs did not know the carriers had left or they would have stayed around and finished the job.

When WWII started, so many new men who were inexperienced in combat joined the Navy that many blunders were happening. The officers and the men did not know what to do next. That is why an officer walked out on the deck and got his head blown off. Leonard said he was so shell-shocked by the experience that he would not have been able to pull a trigger if his life depended on it.

I told Leonard about my rocking the ship experience and asked him if he had any stories in which they had done anything like that. He said that he did. After his rescue, he was taken to the island, New Caledonia, east of Australia so they could recover from their wounds. He said it must have been around Christmas 1942. He went out one night and got so drunk he did not remember much the next day. That night he climbed the signal tower and tapped out the ID signal to see if any ships were out in the sea. It turns out there was an entire US fleet of ships out there. When they saw the signal, they all responded with the secret code to identify them as U.S. ships. Leonard then proceeded to tap out on the signal light, "Merry Christmas." Then men on the ships replied with, "Happy New Year." Now, all of this would seem harmless, except all ships were supposed to be in complete darkness and silent so the Japs would not find them. There were Jap subs in the area and they were looking for U.S. ships to sink. The next morning a meeting was held, and they wanted to know who sent the message to all the ships. Leonard was so drunk the night before that he had no recollection of doing it. Leonard started to say, "What kind of idiot would do that?" when his buddy told him to shut up and be quiet before he talked himself into trouble.

I asked Leonard if he had attended any ship reunions. He said he attended three over the last 71 years. As of this writing, only he and two other men out of the 430 survivors are still living. I asked him if he had had an opportunity to visit the WWII memorial in Washington DC. He replied with a proud, "YES."

Leonard and I were 30 years apart in age and service in the Navy, but we had so many similar experiences. He was there in WWII and I in Vietnam. He said to me, "Mine was not to ask why, but to do or die." That same phrase was being touted during my years in the Navy.

I told Leonard about my meeting with LT Cmdr. Charlie Plumb and asked him whether he would rather have his ship sunk from underneath him or being a POW for six years? We both agreed that no matter how bad your circumstance, you could always find someone who may have it a lot worse than yourself. Leonard said the things that are hard for us are the way by which we learn our wisdom. He said, "Touching a hot stove teaches us we are burnt from hot things."

I have made a friend that will last into eternity. I told Leonard that I feel we were meant to meet. When I went into the cave, he was not there, but just seconds later, he was standing right behind me. I did not see him arrive. I told him we had driven right by that cave and did not stop, but we circled back around and stopped. He told me, he did not know why they were there or what made them stop. He said he had been in that cave hundreds of times in his life and he had no reason to go there again. Leonard agreed with "My final conclusion to all this…" as written on the last page.

Returning To Great Lakes

In June 2012 I returned to Great Lakes Naval Station Training center to visit. I was not even allowed through the gate. I spent 10 months of my life calling it home, and, now, because of terrorists I cannot come through the gate. In 1966 my dad not only went in, but he was able to go wherever he wanted. How things have changed. We have given up our freedom to people in another country, and I am sure they are laughing about how they are controlling us. When a man who once defended this country is now considered a threat to this country, things are totally out of control. We could put our country back the way it was by stopping people who are a potential threat to our way of life from getting into or staying in our country.

The Great Lakes Naval Station now has a very informative museum about US Navy history. A replica of the airplane that Senator John McCain was flying when he was shot down over North Vietnam is on display in the courtyard. The museum sits outside the main gate and is open to the public.

Finding Old Friends in 2013

In researching information for this book on the internet, I found a website www.navybuddies.com and have since made contact with a few shipmates, Barry Rittle, Allan Ebell, Sonny Richardson, Bill Hollywood, John Hand, Jack White, Ron Smith, Allen Grant, Roy Moruss, and Ron Brown. The consensus among many of my fellow sailors is we did not want to be there, but we would make the most of it. Now we are all proud of our service.

After the Navy where did they go? Mike Lopez worked for Anaconda Copper Mines in Arizona. Barry Rittle became an electrician for the nuclear power plant at Three Mile Island in Pennsylvania. Ron Brown worked for McDonald Aircraft in St. Louis, Missouri.

Steve Spratt delivered bulk fuel to farmers in Illinois. LT Jim Brown became a lawyer in Ohio. John Hand became an architect in Georgia. I became a computer programmer in Iowa. Lynn VanAcker became a heavy equipment operator in Illinois. Leonard Joslin became a meat deliveryman in Iowa. Bill Hollywood became a maintenance mechanic at a government building in Alaska. Wayne Maddy became a church pastor in Ohio and John Maddy became a business entrepreneur in Kentucky. Sonny Richardson worked in the oil refinery business for 20 years and for the past 20 years drove a school bus in Indiana. Jack White started an auto body repair shop that he ran until a back injury prohibited from doing it any longer, then he spent the rest of his working career as a quality control inspector for a nuclear manufacturing plant in Connecticut. Ron Smith became an insurance agent for Farmers insurance in Oregon. Roy Moruss worked for Pacific Telephone Company as a communications technician in California. Paul Maddy became a truck driver in Illinois. Allen Grant worked in the electronic field for a while, then started a landscaping business and later managed a swimming pool supply store in Florida. Charlie Plumb became a motivational speaker.

In July 2013, I stopped back at the old homestead, and the door I had put the hammer through was still there. It had been repaired, but where the wood had been cracked was obvious.

The Evening Prayer

As I mentioned in Chapter Six under "Turning In," the Chaplin said an evening prayer every night. Even though at that time in my life I was not sure I believed in God anymore, I still found it comforting to hear his prayer.

While doing my research for this book I discovered the group "Freedom from Religion Foundation" has been trying to get the practice stopped.

> On behalf of military members who strongly object to official prayers on board U.S. Navy ships, the Freedom from Religion Foundation has drafted and sent a letter, co-signed by several other groups, to Secretary of the Navy Ray Mabus, to stop the unconstitutional prayers.

I find this disgusting. The prayer was non-denominational, and the guys who were non-religious just ignored it. Never once did I hear anyone talk negative about it. I say if you are not a believer, stay out of the Navy. It is now an all-volunteer Navy, and no one is forced to be there. This practice has been a tradition since the Navy started. I am not ready to give it up because someone is offended by it.

If we take the Chaplin and religion out of the Navy, who is going to be there for people like me when my brother is killed in the war? Who is going to perform the memorial service when 77 men on the Frank E. Evans die? We gave Osama Bin Laden a religious ceremony for crying out loud. Are we going to have an atheist perform those duties? Not if I can help it, we are not. The Navy's Chaplains corps has been around for 238 years and now people want to dismantle it. I don't think so. It was approved by the Second Continental Congress in October 1775. The second article of the Navy regulations of 1775 : "The Commanders of the ships of the thirteen United Colonies, are to take care that divine service be performed twice a day on board, and a sermon preached on Sundays, unless bad weather or other extraordinary accidents prevent."

We Serve With Honor

> Tradition, Valor, and victory are the Navy's heritage from the past. To these may be added dedication, discipline, and vigilance as the watchwords of the present and the future. At home or on distant stations, we serve with pride, confident in the respect of our country, our shipmates, and our families. Our responsibilities sober us; our adversaries strengthen us. Service to God and country is our special privilege. We serve with honor.

Those words were written on the back of my liberty pass.

My Heroes

After all these years, my Heroes are not Superman or NFL quarterbacks like Brett Farve, but the Men and Women of the United States Armed Forces. It does not matter how big or small your role is/was when you are in the military, you are necessary to the overall objective. The man who pulls the trigger cannot do it if the man in the supply line has not brought him the bullets. Less than 8% of all Americans have served in the military. In fact, everyone in the military does deserve the…

National Defense Service Medal

…just for showing up.

Wear it with pride!

How Ridiculous!

How ridiculous wars are! My mother was right after all. Wars are the result of an egomaniac who wants to control other people. He wants to control either their money, land, religion, or their lifestyle, or all these things. People don't want to be controlled by someone else. They want the freedom to choose their own way. The egomaniac kills thousands or millions of people who disagree with his way of thinking. People fight back from this control freak and hence, we have wars. In the end the egomanic dies, too, either from the sword or old age. Adolf Hitler, Mao Zedong, Sadam Hussein, Osama bin Laden, Ho Chi Min and soon Fidel Castro, all dead. What was it all for? In the big scheme of things the egomaniac was just a blip in time who left the world worse than he found it. Unfortunatly, these people have enough charisma to recruit people to follow them and do their dirty work. We need a military to defend freedom from this kind of person. Hooray for these brave men and women who stand up and fight against tyranny.

My Final Conclusion To All This Is

Everyone who crosses your path during your life's journey can have either a positive influence on you, or they can have a negative influence on you. You can impact their life as well. If years after you met them, they reflect back on that meeting and say, "I really liked that person," you had a positive influence. If they reflect back and see you as I see Chief I. M. Sheethead, your influence was a negative one. If they look at you as I see Lawrence Willey and Glen Johnson, your influence was a positive one. How do you want to be remembered?

About being in the Navy, I have always said this…

"It was a Million Dollar Experience;
I wouldn't give a nickle to do again."

"GOD BLESS AMERICA"

Bibliography

Chapter 1

1. According to the United States Flag Code, the Pledge of Allegiance "Title 4, Chapter 1, Section 4, US Code"
2. Bible Matthew 6:9–13
3. http://www.scoutsongs.com/lyrics/starspangledbanner.html

Chapter 2

1. http://www.militaryauthority.com/wiki/swearing-in-the-oath-of-enlistment-and-the-oath-of-office.html
2. Navy Blue Jackets Manual, chapter 12
3. http://en.wikipedia.org/wiki/Doris_Miller.

Chapter 4

1. Kearsarge name, http://en.wikipedia.org/wiki/Mount_Kearsarge_%28Merrimack_County,_New_Hampshire%29
2. Transit of Venus, http://en.wikipedia.org/wiki/Transit_of_Venus,_2012
3. Kearsarge sloop-of-war, http://en.wikipedia.org/wiki/USS_Kearsarge_%281861%29
4. Kearsarge battleship BB-5, http://en.wikipedia.org/wiki/USS_Kearsarge_%28BB-5%29
5. Flagship description, https://en.wikipedia.org/wiki/Flagship
6. http://en.wikipedia.org/wiki/USS_Kearsarge_%28BB-5%29
7. http://en.wikipedia.org/wiki/USS_Kearsarge_%28BB-5%29
8. http://en.wikipedia.org/wiki/USS_Sailfish_%28SS-192%29
9. http://en.wikipedia.org/wiki/USS_Kearsarge_%28CV-33%29
10. https://en.wikipedia.org/wiki/Essex-class_aircraft_carrier
11. http://Navysite.de/cv/cv33.htm
12. Ships Log in Korea, http://www.angelfire.com/wi2/cvs33/cv.html
13. http://Navysite.de/cv/cv33.htm#history
14. http://Navysite.de/cv/cv33.htm#history
15. Russian rescue, http://www.history.Navy.mil/wars/kearsage.htm
16. Russian Sub in Columbia River, http://www.angelfire.com/wi2/cvs33/cv.html
17. http://www.nzmaritime.co.nz/oriana.htm
18. Space program, http://en.wikipedia.org/wiki/USS_Kearsarge_%28CV-33%29
19. Space program, http://en.wikipedia.org/wiki/USS_Kearsarge_%28CV-33%29
20. 1964 Vietnam Cruise, http://en.wikipedia.org/wiki/USS_Kearsarge_%28CV-33%29
21. Local newspaper

Chapter 5

1. Kearsarge hand out on ship facts from Sonny Richardson

Chapter 6

1. http://www.Navyformoms.com/group/airman/forum/topics/what-the-shirt-colors-mean-on
2. Kearsarge hand out on ship facts from Sonny Richardson

Chapter 7

1. https://en.wikipedia.org/wiki/Mount_Fuji
2. http://www.history.Navy.mil/photos/events/wwii-pac/misc-42/dooltl.htm

Chapter 8

1. Bing Crosby – Don't' Fence Me In Lyrics, http://www.lyrics007.com/Bing%20Crosby%20Lyrics/Don%27t%20Fence%20Me%20In%20Lyrics.html

Chapter 10

1. Sonny Richardson's report to the Captain
2. http://en.wikipedia.org/wiki/Atomic_bombings_of_Hiroshima_and_Nagasaki

Chapter 11

1. https://en.wikipedia.org/wiki/USS_Pueblo_%28AGER-2%29

Chapter 12

1. http://en.wikipedia.org/wiki/Laser
2. http://science.nasa.gov/science-news/science-at-nasa/2004/21jul_llr/

Chapter 15

1. http://en.wikipedia.org/wiki/USS_Kearsarge_%28LHD-3%29

Chapter 16

1. I'm No Hero by Charlie Plum, Independence press, chapter 3, page38
2. HMAS Melbourne, http://en.wikipedia.org/wiki/HMAS_Melbourne_%28R21%29
3. USS Quincy, http://en.wikipedia.org/wiki/USS_Quincy_%28CA-39%29

Picture Credits

Chapter 1

1. Dale, Larry, and Wayne photo in 1952, family snapshot photo
2. RETS business card, scanned
3. Dale's, high school Graduation photo 1965
4. John Deere Part Depot, photo by Dale Maddy 1972
5. Robert Bachman business card, scanned
6. Dale on his Honda 250 motor cycle, Dale's snapshot photo 1966
7. Dale's 1958 Studebaker, photo by Dale Maddy 1966

Chapter 2

1. Lynn VanAcker, high school Graduation photo 1966
2. Dale washing his clothes in boot camp, Dale's snapshot photo 1966
3. Dale right after firefighting class in boot camp, Dales snapshot photo 1966
4. Dale's locker in boot camp, photo by Dale Maddy 1966
5. Picture of Doris Miller Official US Navy Photo. Picture taken from US National Archives 1941 http://www.history.Navy.mil/photos/pers-us/uspers-m/d-miller.htm - "All information on this site is in the public domain and may be distributed
6. National Defense Service Medal, scanned.
7. Larry Maddy, high school Graduation photo 1964
8. Steve Spratt, high school Graduation photo 1966
9. Boot camp Company 536 Graduation photo, by US Navy. 1966

Chapter 3

1. Wayne, Larry, and Dale at Christmas 1966, family snapshot photo.
2. Allan Ebell, Craig Howell and Dale in June 1967, family snapshot photo.
3. Diamond pin, , photo by Dale Maddy 2013
4. Lynn VanAcker, Virginia Gibson, Dale, Lynn Morris, Allan Ebell, and Marie Spitz, Dale's snapshot photo 1967
5. Navy class "A" school graduation photo, by US Navy 1967

Chapter 4

1. Kearsarge mountain This file is licensed under the Creative Commons Attribution 3.0 Unported license.
2. Sloop-of-war Kearsarge, http://www.google.com/imgres?imgurl=http://www.starbase400.org/intrepid/USS_Kearsarge_%281861%29.jpg&imgrefurl=http://www.starbase400.org/intrepid/Intrepid-histo-ry.html&h=167&w=301&sz=1&tbnid=HwSJJMVaSD0aRM:&tbnh=160&tbnw=288&zoom=1&usg=__Wj7rQMe1mQU9cum1Oj1gjAfS44c=&docid=vVAh3zeC6j9AMM&itg=1&sa=X&ei=2drIUZfjOYuBqwHMmYDADA&ved=0CIUBEPwdMA4
3. USS Kearsarge BB-5 source: http://www.navsource.org/archives/01/05b.htm This file is licensed under the Creative Commons Attribution 3.0 Unported license
4. USS Kearsarge AB-1 source: USN photo This file is licensed under the Creative Commons Attribution 3.0 Unported license
5. USS Squalus https://www.google.com/search?q=squalus+submarine&client=firefox-a&hs=hee&rls=org.mozilla:en-US:official&tbm=isch&tbo=u&source=univ&sa=X&ei=C_nIUbXqDpGqqQGjqIC4CA&ved=0CC0QsAQ&biw=1920&bih=961#facrc=_&imgdii=_&imgrc=NPlH1Pp0QzPLaM%3A%3BxqTOgZjMIqCzSM%3Bhttp%253A%252F%252Fwww.hampton.lib.nh.us%252Fhampton%252Fhistory%252Fships%252Fusssqualus%252Fsq

ualus.jpg%3Bhttp%253A%252F%252Fwww.hampton.lib.nh.us%252Fhampton%252Fhistory%252Fships%252Fusssqualus%252Flossofsqualus.htm%3B352%3B321
6. Kearsarge in New York 1945, source: USN photo
7. Kearsarge in Mediterranean 1948, source: USN photo
8. Kearsarge 1952, source: USN photo
9. Korean war ribbon, scanned
10. F2H Banshee jet, source: USN photo 1951
11. F9F Panther Jet, source: USN photo 1951
12. AD4 Sky Raider, source: USN photo 1951
13. F4U Corsair, source: USN photo 1952
14. Bruce Simonds, http://www.koreanwar.org/html/korean_war_project_remembrance.html
15. Engaged barrier landing, source: USN photo 1952
16. Kearsarge in Hong Kong, source: USN photo 1952
17. Denny Phillips, http://www.koreanwar.org/html/korean_war_project_remembrance.htmlhttp://www.koreanwar.org/html/korean_war_project_remembrance.html
18. Donald Hugo Hagge,
19. Kearsarge with hurricane bow 1956, source: USN photo
20. Kearsarge in San Francisco 1957, source: USN photo
21. Oriana, http://www.nzmaritime.co.nz/oriana.htm
22. Map of mercury landings- Google earth
23. Sigma 7 landing 1962, source: USN photo
24. Kearsarge in Hawaii, source: USN photo 1963
25. Vietnam war ribbon, scanned

Chapter 5
1. Entering Pearl harbor, photo by Dale Maddy 1967
2. USS Arizona, photo by Dale Maddy 1967
3. Dr. Robert Maehara, photo by Dale Maddy 1967
4. Don Ho, photo by Dale Maddy 1967

Chapter 6
1. On the flight deck, photo by Dale Maddy 1967
2. Airplane falling off the flight deck, photo by Dale Maddy 1967
3. Ship stores at sea transfer, photo by Dale Maddy 1967
4. Oil tanker transferring fuel, photo by Dale Maddy 1967
5. U.S. Military Payment Currency, scanned
6. Japanese YEN, scanned
7. Philippine Pesos, scanned

Chapter 7
1. Mount Fuji in Japan, photo by Sonny Richardson 1967
2. Flag on fantail, photo by Roy Moruss 1967
3. Stop action shot of helicopter, photo by Dale Maddy 1967

Chapter 8
1. Dragon wrist band, scanned
2. Kowloon apartment buildings, photo by Dale Maddy 1968
3. Walled city, photo by Dale Maddy 1968
4. Boat city, photo by Dale Maddy 1968

Chapter 9
1. Cubi point, Cubi Point photographer 1967
2. Olongapo home, photo by Dale Maddy 1967
3. Salamander lizards, photo by Dale Maddy 1967
4. Banca boats, photo by Dale Maddy 1967
5. Pagsanjan Falls, photo by Dale Maddy 1967

6. Kearsarge from Binictican Golf course, photo by Dale Maddy 1967

Chapter 10

1. Kearsarge 1968, photo by Dale Maddy 1968
2. Nagasaki ground zero monument, photo by Dale Maddy 1967

Chapter 11

1. Armed Forces Expeditionary Service, scanned
2. 1967-68 cruise, Google earth map image

Chapter 12

1 USS Kearsarge patch, scanned
1. Row boat, photo by Dale Maddy 1968
2. Meritorious Unit Commendation, scanned
3. Captain Cook, source: USN photo 1968
4. Kearsarge patch, scanned 2013
5. Calibration school graduation photo, by US Navy 1968
6. Battle Efficiency patch, scanned 2013

Chapter 13

1. LT Larry Maddy, family snapshot photo 1968
2. Larry's Funeral, *Times Democrat Newspaper*, Davenport, Iowa 1969

Chapter 14

1. Tora! Tora! Tora! movie, photo by Dale Maddy 1969
2. Gary Howlett and Larry Maddy, family snapshot photo 1966
3. The Thinker, photo by Dale Maddy 2013
4. Cubi Point group photo, by a local Pilipino photographer
5. Landing boat, photo by Bill Hollywood, Ketchikan, Alaska 1969
6. USS Frank E Evans, photo by Dale Maddy 1969
7. HMAS Melbourne, photo by Dale Maddy 1969
8. Moon landing, photo by Dale Maddy 1969
9. Apollo 11 recovery map, Google earth image
10. At sea explosion, photo by Bill Hollywood Ketchikan, Alaska 1969
11. Sitting on the Front of the ship, photo by Dale Maddy 1969
12. Swim call, photo by Bill Hollywood, Ketchikan, Alaska 1969
13. Group during a Chinese dinner, photo by Dale Maddy 1969
14. Victoria bay in Hong Kong, photo by Dale Maddy 1969
15. Rickshaw drive, photo by Dale Maddy 1969

Chapter 15

1. Dale's 1969 Mercury Cougar, photo by Dale Maddy 1970
2. Dale on last day in the Navy, Dales snapshot photo 1970
3. Kearsarge decommission photo, source: USN photo 1970
4. USS Kearsarge LDH-3, source: USN photo 1993

Chapter 16

1. Connie Francis, photo by Dale Maddy 1967
2. LT Cdr Charlie Plumb, photo by Fletcher 1973
3. Flag of Honor, photo by Dale Maddy 1992
4. Vietnam wall, photo by Dale Maddy 2001
5. Taos Pueblo, photo by Dale Maddy 2007
6. Willy Fudd, photo by Dale Maddy 1968
7. Carrier landing approach, photo by LT Jim Brown 1968
8. USS Quincy, source: USN photo 1940
9. USS Quincy burning, photo by Japanese sailor 1942
10. Leonard Joslin at 25 years old, Navy photo
11. Dale and Leonard Joslin, photo by Tina Bryant 2011
12. Leonard's autograph, scanned 2011
13. National Defense Service Medal, scanned

14. Kearsarge photo 1967, source: USN photo

People Index

About The Author

1947 - Born in Belleville, OH
1948 - Family moved to Granville, IL
1953 - Attended grade school in Granville, IL
1955 - Attended grade school in Standard, IL
1958 - Family moved to Sheffield, IL
1961 - Student, Western High School in Buda, IL
1965 - Graduated from high school
1965 - Student, RETS school in Rock Island, IL
1965 - Laborer, Barcol Overdoors in Sheffield, IL
1965 - Parts picker, John Deere in East Moline, IL
1966 - Joined United States Navy
1967 - Boarded the USS Kearsarge in Long Beach, CA
1970 - Parts picker, John Deere in East Moline, IL
1971 - Part time free-lance wedding photographer
1971 - Parts packer, John Deere in East Moline, IL
1972 - Computer analyst, John Deere in East Moline, IL
1976 - Area Sales Manager, Agri-King, Inc. in Maquoketa, IA
1977 - Regional Sales Manager, Agri-King, Inc. in Maquoketa, IA
1980 - Part time magician entertaining friends and family
1983 - Division Sales Manager, Agri-King, Inc. in IL and IA
1985 - Vice President, Maddy Nutrition, Corp. in Clinton, IA
1985 - Student, Clinton Community College in Clinton, IA
1987 - Owner, PC BusinessWare, computer consulting in IA
1988 - President, Taking Care of Business, Inc. in IA
1991 - Executive Vice President, ANC, Inc. in WI
2010 - Retired
2013 - Author